AMERICAN CHEMICAL SOCIETY

Safety in Academic Chemistry Laboratories

8TH EDITION

BEST PRACTICES FOR FIRST- AND SECOND-YEAR UNIVERSITY STUDENTS

Foreword from the Chair

At a time when there was little emphasis on teaching laboratory safety, the Committee on Chemical Safety of the American Chemical Society (ACS) published the first edition of *Safety in Academic Chemistry Laboratories (SACL)*. Now, more than four decades later, *SACL* has undergone seven revisions, is printed in three languages, and is one of the most widely used laboratory safety guidance documents in print and online. Although it was initially written for academic chemistry laboratories, its information and application of safe practices can be extended to all laboratories, research facilities, and workplaces where chemicals are used.

Over time, we have become more aware of the hazards and risks associated with laboratory chemicals and have sought to share this new information. The Committee,

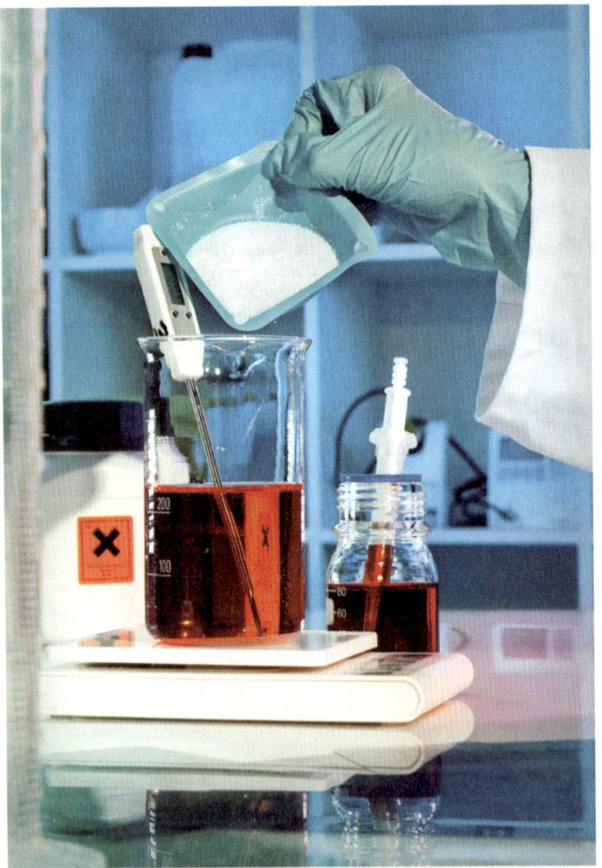

along with the ACS Division of Chemical Health and Safety, disseminates current materials, practices, and research developments through technical programming at national and regional meetings, articles in its *Journal of Chemical Health and Safety*, and publishing guidance documents for academic and industrial institutions. After several academic laboratory accidents, the Committee recognized a great need for improved safety education in all academic environments. A task force generated *Creating Safety Cultures in Academic Institutions*,[1] a guidance document designed to improve the safety culture in these institutions. To assist academia in strengthening the education aspect of safety, the Committee generated several additional guidance documents. *Guidelines for Chemical Laboratory Safety in Academic Institutions*[2] integrates safety education throughout the entire curriculum for all levels of chemistry using a novel acronym system. Precollege (high school and middle school) science educators include chemical safety in the curriculum by following *Guidelines for Chemical Laboratory Safety in Secondary Schools*.[3] The Committee also provides two publications to assist elementary science teachers with safety education. Copies of *Safety in the Elementary (K–6) Science Classroom* and *Chemical Safety for Teachers and Their Supervisors: Grades 7–12* are available through the ACS by request at safety@acs.org.[4] Additional safety resources are available through download at www.acs.org/safety.

On behalf of the Committee on Chemical Safety, I am pleased to introduce this eighth edition of *SACL*. I thank all the contributors who generously contributed their time, talents, and expertise to this and previous editions of this outstanding publication. Their efforts make the academic laboratory a safer place in which to work and learn. A special thank you goes to David C. Finster, who served as Editor and chief contributor for this edition (*SACL-8*). The comprehensive revisions and new sections are a direct result of his commitment and dedication to chemical safety education. The Preface from the Editor acknowledges those who contributed to this edition and previous revisions. Marta Gmurczyk coordinated the ACS staff members involved in the production and distribution of *SACL-8*.

All comments are welcome. Please direct them to the ACS Committee on Chemical Safety at safety@acs.org.

Elizabeth M. Howson
Chair, ACS Committee on Chemical Safety
March 2017

---

[1] ACS Joint Board/Council Committee on Chemical Safety. *Creating Safety Cultures in Academic Institutions: A Report of the Safety Culture Task Force of the ACS Committee on Chemical Safety*; American Chemical Society: Washington, DC, 2012. www.acs.org/content/dam/acsorg/about/governance/committees/chemicalsafety/academic-safety-culture-report-final-v2.pdf

[2] ACS Committee on Chemical Safety. *Guidelines for Chemical Laboratory Safety in Academic Institutions*; American Chemical Society: Washington, DC, 2016. www.acs.org/content/dam/acsorg/about/governance/committees/chemicalsafety/ publications/acs-safety-guidelines-academic.pdf?logActivity=true

[3] ACS Committee on Chemical Safety. *Guidelines for Chemical Laboratory Safety in Secondary Schools*; American Chemical Society: Washington, DC, 2016. www.acs.org/content/dam/acsorg/about/governance/ committees/chemicalsafety/publications/acs-secondary-safety-guidelines.pdf?logActivity=true

[4] ACS Committee on Chemical Safety. Chemical Safety in the Classroom. www.acs.org/content/acs/en/about/governance/committees/chemicalsafety/chemical-safety-in-the-classroom.html

Preface from the Editor

The first edition of *Safety in Academic Chemistry Laboratories* (*SACL*) was written in 1972 by members of the ACS Committee on Chemical Safety (CCS) under the direction and urging of its chair, Howard H. Fawcett. It was published as an 11-page, double-spaced, typed and mimeographed document. Since then, more than a million copies of the first seven editions of *SACL* have been distributed. Throughout the intervening years, the field of chemical health and safety has evolved considerably, although some basic concepts remain unchanged. This current edition represents the current best practices in academic laboratory safety.

The purpose of this booklet (*SACL-8*) is to provide an overview of the key issues related to the safe use of chemicals in the first two years of an undergraduate program in chemistry. Some information of a more advanced nature has been pruned, because delving into these topics would have considerably increased the size of the booklet

and because the primary use of this booklet has been and continues to be for topics appropriate for first- and second-year college students. This change is reflected in the title, along with the focus on a positive safety culture with the phrase "best practices".

Much from previous editions has been retained. Some changes from *SACL-7* include:

- A new title, to reflect that the booklet has been revised to target safety concerns encountered by first- and second-year college students in general chemistry and organic chemistry laboratories.
- A new introduction that sets basic laboratory safety information in the context of developing a culture of safety in chemical laboratories.
- A review of common personal protective equipment and various safety practices in laboratories.
- A guide to chemical hazards, how to recognize them, and sources of information about chemical hazards, including the GHS.
- An overview of safety concerns for many common laboratory techniques.
- An overview of safety equipment and emergency response procedures for fires, spills, and chemical exposures.
- The addition of sidebars, for interest and readability, and "In Your Future" sections for common laboratories hazards and concerns that are less likely to be encountered in introductory and organic laboratories but about which all students should be aware.
- A content shift from safety based primarily on rules to learning about safety through the RAMP principles.[1]

The list of individuals who have contributed to *SACL-8*, and all previous editions, is too long to include here in its entirety. It is appropriate, however, to note the contributions of Jay Young (1920–2011) over the span of decades, including the first edition. This is but one of his many long-lasting contributions to the world of chemical safety. For this edition, the primary authors have been Georgia Arbuckle-Keil, Robert H. Hill, Jr., Samuella Sigmann, Weslene Tallmadge, and myself. Debbie Decker, Harry Elston, and Ed Movitz provided technical reviews. Thanks to Karen Müller, who provided editorial services, Amy Phifer of Plum Creative Services, who designed this new edition of the publication, and Marta Gmurczyk (ACS staff liaison to the CCS), who coordinated the ACS staff members involved in the production and distribution of SACL-8.

David C. Finster
Editor
March 2017

---

[1] Hill R. H., Finster D. C. *Laboratory Safety for Chemistry Students*, 2nd ed.; John Wiley & Sons: Hoboken, NJ, 2016.

Disclaimer

The materials contained in this manual have been compiled by recognized authorities from sources believed to be reliable and to represent the best opinions on the subject. This manual is intended to serve only as a starting point for good practices and does not purport to specify minimal legal standards or to represent the policy of the American Chemical Society. No warranty, guarantee, or representation is made by the American Chemical Society as to the accuracy or sufficiency of the information contained herein, and the Society assumes no responsibility in connection therewith. This manual is intended to provide basic guidelines for accident prevention. Therefore, it cannot be assumed that all necessary warning and precautionary measures are contained in this document and that other or additional information or measures may not be required. Users of this manual should consult pertinent local, state, and federal laws and legal counsel prior to initiating any accident-prevention program.

Registered names and trademarks, etc., used in this publication, even without specific indication thereof, are not to be considered unprotected by laws.

Safety in Academic Chemistry Laboratories
BEST PRACTICES FOR FIRST- AND SECOND-YEAR UNIVERSITY STUDENTS

Table of Contents

CHAPTER 4 Recommended Laboratory Techniques

CHAPTER 5 Safety Equipment and Emergency Response

CHAPTER 1

Being Safe in the Laboratory

Introduction

Chemistry laboratories present more hazards than are typically found in other science laboratories. Interestingly, the very properties that we value in some chemicals are also what make them hazardous. For example, we like the fact that some organic solvents dissolve organic molecules very nicely, but this same feature also makes them dry out our skin. We use the reactivity that acids and bases provide in order to effect a chemical change, but that reactivity also makes them hazardous if they are in contact with skin or are ingested. We like the fact

that liquid nitrogen and dry ice are both *very* cold, because sometimes we need very low temperatures, but these substances are dangerous to handle with bare hands precisely because they are so cold. We like the fact that natural gas burns (in a Bunsen burner), but if it builds up in an enclosed space, a spark or flame can cause an explosion.

In any laboratory—a chemistry laboratory or other science laboratory—where chemicals are used, there will be hazards. Well-educated chemists and well-educated chemistry students need to understand the hazards of chemicals and of various chemical procedures in order to work safely in the laboratory. For the ability to work safely, it is key to not only recognize a hazard but also assess the actual risk it poses. For example, an organic solvent might be very flammable, but the risk in a particular laboratory is low if the solvent is well contained in a bottle or if the amount of solvent is very small and sources of ignition are excluded from the area. Similarly, a chemical might be very toxic by ingestion, but if we avoid conditions where it would be ingested, the risk is low. Finally, a strong acid might be very corrosive to the skin, but if we take steps to avoid skin contact, the risk is lower. *In chemistry laboratories, there are always some hazards and some risks.* One goal of this booklet is to help you learn how to recognize hazards and how to minimize risks.

Your instructor is requiring you to read this booklet so that you can work safely in laboratories with chemicals. Conducting chemistry experiments can be fun, intellectually satisfying, and productive – but only if these experiments are conducted as safely as possible. At the end of each day in the laboratory, the goal is for you to go home just as you came – with no injuries or illnesses as a result of your laboratory experience.

In order to work safely in your first laboratory courses, you will learn lots of rules about chemicals and how to handle them. Indeed, this booklet is full of rules, too. But there are *principles of safety* (RAMP) behind all of these rules, and learning these principles is also a goal of this booklet. The rules in this booklet deal with many situations that you will encounter in first- and second-year chemistry courses. But there are many additional hazards that you may encounter in more advanced courses and in research projects. So, while learning some rules, you should also start to develop principles and concepts about safety that can be applied in your future working life, even though you might not be working specifically in a chemistry laboratory. For non-chemistry science majors, your "non-chemistry" laboratories will often use chemicals. As the saying goes, "The chemicals don't know what laboratory they are in." Many chemicals are used in the home also, so handling and using chemicals safely has a broad range of application.

Safety in Academic Chemistry Laboratories is designed for use as an aid in teaching safety during the first two years of chemistry courses. In these early years, most laboratory experiments will have been carefully reviewed for safety by your institution's chemistry faculty. The risks of these experiments have been minimized, and there is considerable oversight of these early laboratory sessions.

Safety Culture and Your Role in It

The safety knowledge and skills that you learn in your chemistry courses is greatly influenced by the safety culture of your institution. The components of a strong safety culture require *you* to do your part. There are four areas that should receive your attention: leadership, learning safety, building a positive safety attitude, and learning lessons from safety incidents. You can learn much more about safety culture by reading *Creating Safety Cultures in Academic Institutions.*[2]

Leadership plays a critical role in the kind of safety culture an institution will have. Although you may have no role in the leadership of your institution, you *do* have a role in being a personal leader in safety. As in other aspects of life, safety is encouraged by good example. You can show your leadership by following the safety instructions given to you by your instructors, by always wearing your required personal protective equipment (such as safety goggles, gloves, and a laboratory coat), by reporting all safety incidents (however minor), and by taking time to consider the risks involved in an experiment.

As you learn chemistry, you will be learning about laboratory and chemical hazards and how to minimize the risks of those hazards in the laboratory — elements of laboratory safety. It is crucially important to your personal safety (and to the safety of others) that you really try to learn about and to understand hazards in the laboratory. This understanding can save you or others from injuries or other adverse incidents. Understanding *why* chemicals or situations are hazardous is a critical part of learning to be safe. For example, if you know *why* something is flammable, you will be better able to work with flammable materials in the future. As you learn chemistry, you should endeavor to learn as much about safety as you can. We hope that you will be learning about safety throughout your entire undergraduate experience and will not view safety as just a set of rules to memorize. A safety education involves building a substantial knowledge of the various chemical and laboratory hazards, learning how to evaluate the risks of those hazards, learning how to minimize the risks of all the hazards you may encounter, and being prepared for emergencies that might arise during laboratory work with these hazards — we're talking about RAMP here.

If you continuously learn about safety during your undergraduate years of education, and its importance is constantly reinforced, you should be building a positive safety attitude, sometimes called a safety ethic. If you always take time to review the hazards and safety measures of each and every experiment, then you will also be maintaining your positive safety attitude. The proper attitude for safety is reflected in *the safety ethic:* value safety, work safely, prevent at-risk behavior, promote safety, and accept responsibility for safety.[3] A positive safety attitude will likely be expected and required by your future employer.

Finally, although we all strive to keep everyone safe in our laboratory operations, adverse safety incidents can sometimes happen. Much of what is known about safety has been learned from mistakes or incidents. When these happen — even minor incidents — they need to be shared so that we can all learn lessons from the missteps that were made. Sharing these incidents should always be done in a nonpunitive way. Learning lessons is an important part of our work as scientists, so keep this in mind as you learn about chemistry and safety.

As you can tell by this introductory chapter, safety plays a key role in chemistry. Safety concerns apply across all chemistry and related fields, and safety is in fact a discipline of chemistry, just like inorganic, organic, analytical, physical, or biological chemistry. Everyone using chemistry in their career needs adequate knowledge, skills, and attitudes about safety to work safely in a laboratory. *Keep safety at the forefront in your chemistry and science education, and it will serve you well.*

SUMMARY

Working in laboratories requires that you learn to apply the RAMP concept: Recognize hazards, Assess the risks of hazards, Minimize the risks of hazards, and Prepare for emergencies. Learning the *why* about hazards, their risks, and procedures and processes designed to protect you is the basis for safety rules. If you understand *why*, you are more likely to follow safe procedures and safety rules.

REFERENCES

[1] Hill, R. H.; Finster, D. C. *Laboratory Safety for Chemistry Students*, 2nd ed.; John Wiley & Sons: Hoboken, NJ, 2016.

[2] ACS Joint Board–Council Committee on Chemical Safety. *Creating Safety Cultures in Academic Institutions: A Report of the Safety Culture Task Force of the ACS Committee on Chemical Safety*; American Chemical Society: Washington, DC, 2012. www.acs.org/content/dam/acsorg/about/governance/committees/chemicalsafety/academic-safety-culture-report-final-v2.pdf (accessed March 6, 2017).

[3] Hill, R. H. The Safety Ethic: Where Can You Get One? *Chem. Health Safety* 2003, *10*, 8–11. http://dx.doi.org/10.1016/S1074-9098(03)00025-X

Your Responsibility for Safety in Laboratories

Introduction

Incident prevention is a collective responsibility, which requires the full cooperation of everyone in the laboratory. The responsibility for safety in the laboratory resides with you, your peers, the instructor, and the institution. Although everyone is responsible for safety in the laboratory, you, the experimenter, can most directly prevent incidents.

Incidents often result from:

- an indifferent attitude toward safety;
- failure to recognize hazards or hazardous situations;
- failure to assess the risks involved in the work being done;
- failure to be alert to your surroundings;
- failure to follow instructions or measures to minimize risks; and
- failure to recognize the limitations of your knowledge and experience.

You can be a victim of a mistake made by you or by someone else. If you are performing a laboratory procedure incorrectly and a classmate points this out to you, be grateful — it could be that he or she has just saved your life. If you observe a classmate making a mistake, let him or her know. In addition, unsafe acts should be reported to your instructor, so that mistakes will not be repeated.

The guidelines in this chapter are provided to help you to develop an awareness of your role in maintaining a safe laboratory environment. Most of these guidelines and rules will be applicable to the introductory laboratory courses you take as an undergraduate student. In some cases, more information is provided to guide you as you advance to more laboratory experiences. At the end of the chapter, you will

■ Events, Incidents, and Accidents

Unexpected and unwanted events sometimes occur in laboratories. In this booklet, we use the term "incident" instead of "accident" so as not to imply that these events occur randomly or inevitably. Virtually all incidents are preventable if the guidelines and safety principles presented in this booklet are followed.

find a summary list of guidelines or safety rules. These guidelines are the result of the application of years of lessons learned and are part of the M of RAMP: <u>M</u>inimize the risks of hazards. As you learn more about chemistry, you will learn to recognize hazards (Chapter 3). But before you enter a laboratory, you need to learn some basics about safety measures and safety equipment. Review the safety measures frequently, as a reminder.

In introductory and organic chemistry laboratory courses, it is unlikely that you will work with substances at pressures other than ambient, or with high-energy materials with the potential for vigorous reactions, such as foaming, overheating, boiling over, or even — if under pressure — small explosions (these are extremely rare occurrences). These types of laboratory experiments require the use of appropriate laboratory hoods, bench shields, chemical splash and impact goggles to protect your eyes, and face shields wide enough and long enough to protect your neck and ears.

Personal Protective Equipment (PPE)

Personal protective equipment (commonly known as PPE) is one of the principal ways of protecting you from harm when you work in the laboratory. It is important that you understand why your instructor will require you to use PPE.

PPE is used to eliminate or minimize exposure to some hazards encountered when working in the chemistry laboratory. PPE includes items designed to protect specific areas of your body, such as your eyes and hands. It commonly includes gloves, eye protection, laboratory coats, and aprons. Don't depend solely on PPE to protect you, because it is often the final barrier between you and exposure.

Hair and Apparel (Dressing for the Laboratory)

Clothing worn in the laboratory should offer your skin basic protection from splashes and spills. Shorts, short skirts, and shirts that expose your midriff will unnecessarily expose your skin to potential spills. It is always prudent to minimize the amount of skin exposed to the laboratory environment. Bulky and loose-fitting clothing is not appropriate in the laboratory. Loose sleeves may knock laboratory items over, be dragged through chemical spills, or present a fire hazard with open flames. Clothing should be made of natural fibers, such as cotton. Your instructor or institution may require you to wear laboratory coats or aprons. Nonflammable, nonporous aprons offer the most satisfactory and the least expensive protection. If you wear a laboratory jacket or coat instead of an apron, it should have snap fasteners rather than buttons, so that it can be readily removed in case of contamination.

In the laboratory, wear shoes with uppers made of leather or polymeric leather substitute that completely cover your feet and toes (closed-toe shoes). This will offer your feet the best protection from spills and dropped items. As you choose your laboratory footwear, keep in mind that the shoes you wear in the laboratory *should not expose the tops of your feet and should offer stability for standing and walking.*

Constrain long hair and loose clothing. Long hair can easily become entangled in equipment, can be exposed to chemicals, or can catch on fire by direct exposure to lit Bunsen burners. The wearing of jewelry, such as rings, bracelets, necklaces, and wristwatches, in the laboratory should be avoided. Jewelry can be damaged by chemical gases and vapors, and from spills. Chemical seepage between the jewelry and the skin can put corrosives in intimate contact with your skin and trap the chemicals there. Jewelry also can catch on equipment, causing injuries.

Eye Protection

Everyone in the laboratory, including visitors, must wear eye protection at all times, even when not performing a chemical operation. Some experiments present splash hazards, which necessitate wearing indirectly vented goggles; for other experiments, safety glasses can suffice. Because it is likely that you will not wish to purchase two forms of eye protection, it is prudent to use the more protective eyewear for variable environments. Thus, goggles rated for chemical splash protection are the preferred eye protection. The chemistry faculty at your institution will assess the risks of the hazards in your laboratory and determine the appropriate type of eye protection for the experiments being performed in their academic laboratories. Normal prescription eyeglasses do not provide appropriate laboratory eye protection against shrapnel from an explosion or splashes of hazardous chemicals. Serious injuries have resulted from the wearing of normal prescription eyewear without chemical splash goggles or safety glasses.

Gloves

Gloves are an important part of personal protection. Your instructor will assess the risks of hazards and will require the use of gloves when appropriate and provide the proper type of gloves. Glove material must be selected based on the chemicals being used. Always check your gloves before each use to ensure the absence of cracks and small holes. To avoid unintentionally

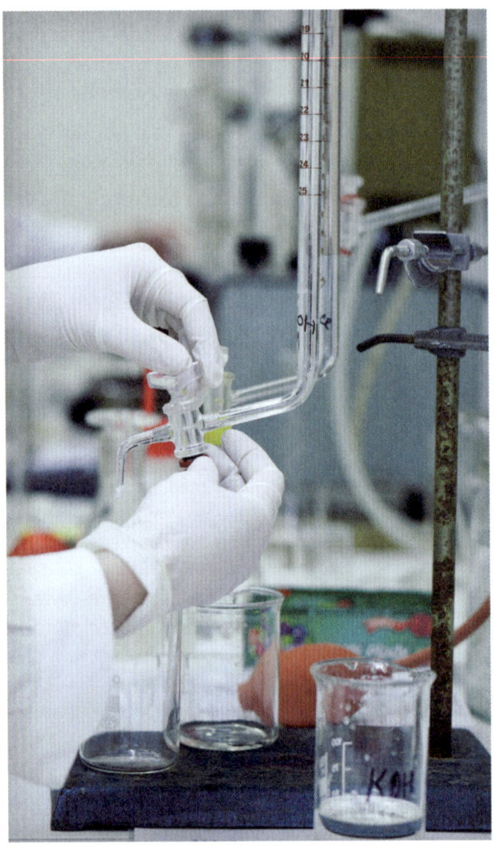

spreading chemicals, remove your gloves before leaving the work area and before handling such things as cell phones, calculators, laptops, doorknobs, writing instruments, laboratory notebooks, and textbooks. You should wash your hands when leaving the laboratory, even if you have worn gloves.

A variety of gloves and materials are available: neoprene, butyl rubber, and many other materials. Different types of gloves have different gauntlet lengths; some cover the entire arm, some cover only the forearm, and some are only wrist-length. Individuals who are latex-sensitive should not wear gloves made of latex. Although cloth or leather gloves may protect against hot or cold objects, do not rely on them for protection against hazardous chemicals. The instructor and the institution are responsible for assessing the risks of hazards and for selecting the proper glove for the particular application.

Disposable gloves and gloves that have been permeated by a chemical should not be reused. The gloves cannot be reused safely because the chemical cannot be totally removed. Contaminated gloves may be considered to be a hazardous waste material, but this is not always the case. In all instances, dispose of your used gloves in the designated hazardous waste container or as directed by your instructor.

■ IN YOUR FUTURE: Selecting Gloves

Be aware that no glove material can provide permanent protection. Eventually, liquids will permeate all glove materials. Glove materials are rated by the manufacturer using the breakthrough time (the time it takes a particular chemical in contact with a glove to pass through the glove). For many organic solvents, the breakthrough time can be only a few minutes. Because the permeability of gloves made of the same material or a similar material can vary by manufacturer, refer to the information provided by the manufacturer of the gloves for specific guidance. If a chemical diffuses through a glove, it is then held against your skin; you could receive more exposure than if you hadn't worn a glove at all. Additional information can be obtained from the manufacturer of the gloves. An online search for "chemical glove selection" will yield several websites with useful information.

Laboratory Protocols

Laboratory Environment

The chemistry laboratory can provide a wealth of opportunity for learning, but while working in the laboratory, you should remain alert to your actions and the actions of those around you. Variations in procedure, including changes in the chemicals to be used or in the amounts that will be used, may be dangerous. Alterations should be made only with the knowledge and approval of your instructor.

Before working in the laboratory, take note of your surroundings. Locate the exits, fire alarm pull stations, eyewash fountains, safety showers, fire blankets, first aid kits, and fire extinguishers; practice walking to them. This is part of the P of RAMP: _Prepare for emergencies.

Never eat or drink in the laboratory, to ensure that there is no chance that any contamination can lead to ingestion of a laboratory chemical. No food or drink should be carried into or stored in the laboratory.

Visitors in the Laboratory

All laboratory visitors, no matter how brief their visit, should wear eye protection. Visitors, such as friends and relatives, may not be aware of the hazards and may inadvertently commit unsafe acts. Obtain your laboratory instructor's approval before bringing visitors into the laboratory.

Housekeeping

In the laboratory and elsewhere, keeping things clean and neat generally leads to a safer environment. Keep aisles and access to safety equipment free of obstructions such as chairs, boxes, open drawers, backpacks, and waste receptacles. Avoid slipping hazards by keeping the floor clear of spilled liquids, ice, stoppers,

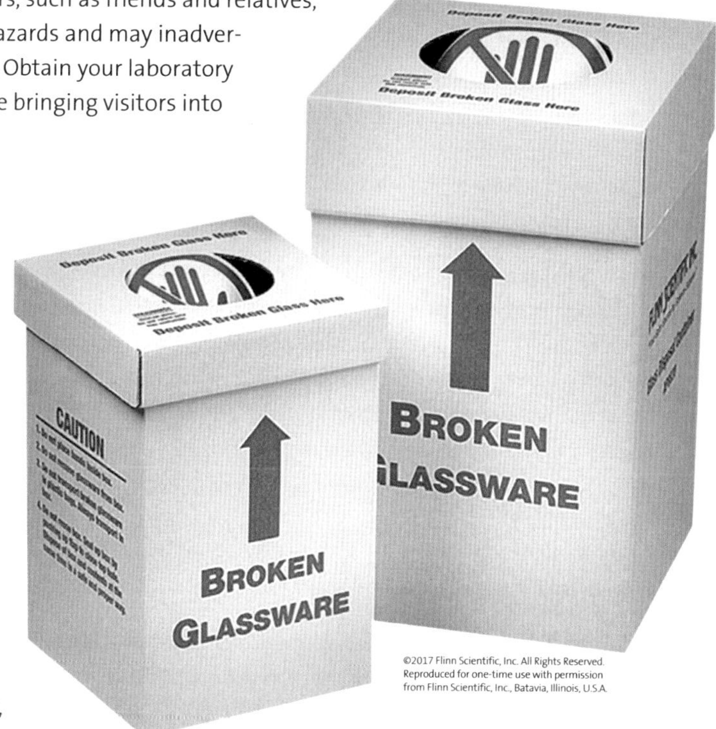

glass beads or rods, and other such small items. Keep workspaces and storage areas clear of broken glassware, leftover chemicals, and dirty glassware. Broken glassware should always be disposed of in a broken glass disposal container and NEVER in an ordinary trash can. Inform your instructor immediately if glass is broken or chemicals are spilled. Follow your laboratory's required procedure for the disposal of chemical wastes and unused chemicals. Wipe your bench area before leaving the laboratory, so that others will not inadvertently touch chemical residue. Never leave chemicals on balances, because this may unnecessarily expose the next user to the chemical; in addition, electronic balances are expensive and can easily be damaged by corrosive chemicals.

Labeling Chemicals

Improper or insufficient labeling of chemical containers has resulted in numerous adverse incidents. Labels are typically referred to as "manufacturer" and "secondary". It is important that a manufacturer label is never altered, covered, or otherwise changed until the container is verified as being empty. Often, empty containers will be reused, for example for solutions prepared by students. Before a container is reused for another solution, the obsolete label should be removed completely and the container should be thoroughly washed and allowed to air dry. It is unacceptable to use a marker to write over an existing manufacturer label. In no instance should a container ever have two labels, one on each side of the bottle.

Although it is unlikely that you will be involved in the management of manufacturer containers, you may prepare solutions and store them in your drawer in your introductory and organic chemistry laboratories. At a minimum, a secondary label for temporary use (during a laboratory period or until a future laboratory period) should have the name of the chemical, the name of the person who filled the container, the date it was filled, and the hazards. Containers prepared for longer storage should have a label that meets the standards of the Globally Harmonized System of Classification and Labelling of Chemicals (GHS) (see Chapter 3).

Cleaning Glassware

Clean your dirty glassware at the laboratory sink using hot water, environmentally acceptable cleaning agents, and brushes of suitable stiffness and size. Do not force a brush into glassware. Always wear chemical splash goggles while washing dishes, and wear gloves if instructed to do so. Many laboratory faucets have serrated nozzles that can produce high-velocity streams of water. When using these types of faucets, always adjust the stream slowly without holding glassware underneath. Water can come out forcefully and splash back into your face or knock the glassware out of your hands. Avoid accumulating too many items in the cleanup area.

Workspace around a sink is usually limited, and piling up dirty or cleaned glassware can lead to breakage. Remember that the turbid water in a sink may hide the sharp, jagged edge of a piece of broken glassware that was intact when it was put into the water. If glassware in the sink is broken, drain out the standing water. Use a pair of cut-resistant gloves, tweezers, or tongs to remove the pieces of broken glass. Be particularly careful when cleaning the drain area, because glass pieces can get caught in the holes and be nearly impossible to spot. To minimize breakage of glassware, sink bottoms may have rubber or plastic mats that do not block the drains.

Inhaling Harmful Chemicals

If you are instructed to smell something in the laboratory, use your hand to waft vapors toward your face and sniff gently. You should never sniff a chemical by placing your nose directly over a chemical container. The presence of an odor is not a reliable indication of potential harm, and the absence of an odor is not a reliable indication of the absence of harm.

Some people think that if they can smell a chemical, it is causing them harm. This is not necessarily correct. It is certainly correct that if you smell a chemical, you are inhaling it. Some harmful chemicals have no odor, and others can paralyze the sense of smell. Some chemicals cannot be detected by the human nose at concentrations that are harmful, and some, even though they might have a decidedly noxious odor, are not harmful if inhaled.

Many substances that may or may not have an odor are harmful if their vapors, dusts, or mists are inhaled. The label on the container and the Safety Data Sheet (SDS) for the chemical (see Chapter 3) may carry a warning about inhalation hazards. Your instructor will direct you to dispense and handle these substances in a laboratory hood.

Disposal of Chemicals

Proper handling of reaction by-products, surplus, waste chemicals, and contaminated materials is a major element of incident prevention, and there are very strict rules for disposing of chemicals. Improper disposal can result in serious damage to the environment and can also result in legal issues for your institution. Every student is responsible for ensuring that these wastes are handled in a manner that minimizes personal hazard and recognizes the potential for environmental contamination.

Typically, your reaction by-products and surplus chemicals will be poured into appropriately labeled waste or hazardous waste containers for proper disposal. Your instructor will direct you to use designated, labeled waste containers. Most likely, different containers will be used for different classes of chemicals. Handle your waste materials in the specific ways designated by your instructor. ***Pouring waste into the wrong container could result in unexpected, adverse reactions, leading to fires or explosions*** *(see "Contains Nitric Acid— DO NOT ADD ORGANIC SOLVENTS" in Chapter 3)*. Remember to pay attention and follow instructions.

Sometimes your reaction by-products can be neutralized or deactivated as part of your procedure, and this can help to reduce waste handling, which lowers the cost of disposal. Once your by-products are moved away from the experiment, they are subject to hazardous waste regulations established by your state government and the federal Environmental Protection Agency (EPA).

Some general disposal guidelines are as follows:

- When disposing of chemicals, put each class of waste chemical into its specifically labeled disposal container. Carefully read the contents label, and replace the cap after use.

- Never put chemicals into a sink or down the drain unless your instructor has told you that local regulations allow these substances to be put into the sanitary sewer system. For example, water and dilute aqueous solutions of sodium chloride, sugar, and soap from a chemistry laboratory may be disposed of in the sink.

- Put ordinary wastepaper into a wastepaper basket separate from chemical wastes. Materials that are contaminated with chemicals, such as paper towels used to clean up a spill, may need to be placed into a special container marked for this use. Your instructor will tell you whether cleanup materials need to be collected for hazardous waste or placed into the landfill containers.

- Broken glass belongs in its own marked waste container. If the broken glass is contaminated with chemicals, ask your instructor where to dispose of the glass. Thermometers that contain mercury may still be in use at your institution, but most of these have been replaced with thermometers that contain alcohol-based liquids. If you happen to be using a mercury thermometer and it breaks, immediately notify your laboratory instructor. Spilled mercury requires special cleanup procedures, and it should not be ignored, because mercury is toxic. Broken thermometers may contain mercury in the fragments; broken glass contaminated with mercury belongs in its own labeled container.

SUMMARY

Most of what we know about science was learned in laboratories somewhere. Laboratories can be interesting places to learn, but they can also be places with hazardous chemicals and equipment. In order to protect yourself and your peers, it is important to be conscious of these hazards and risks and to avoid actions that may lead to incidents that cause injury to you and your classmates or damage to the laboratory. Your duty as a student includes the duty to prevent incidents whenever you are in the chemistry laboratory. The following list summarizes the basic guidelines (to minimize the risks of hazards) intended to help you fulfill this important responsibility. Whenever you are in the laboratory·

Photo courtesy of CP Lab Safety.

PROPER CONDUCT/BEHAVIOR

- Do not work alone.
- Never perform unauthorized experiments or change procedures without approval.
- Maintain an awareness of your surroundings, and move purposefully around others.
- Never remove chemicals from the laboratory without proper authorization, and report to your instructor any observed unauthorized removal of chemicals by others.
- Never play tricks or engage in horseplay in a chemistry laboratory.
- Notify your instructor if you observe violations of your laboratory's safety rules; you could save someone's life.

PROPER LABORATORY ATTIRE

- Prevent skin exposure by covering your skin.
- Feet must be completely covered, and no skin should be showing between the top of the shoe and the bottom of the skirt or pants.
- Confine long hair, avoid wearing loose clothing, and remove scarves and jewelry.

SAFE HANDLING OF CHEMICALS

- Read the procedure ahead of time, listen carefully to your instructor's directions, and note any safety requirements for the experiment in your prelab notes.
- Never directly sniff a chemical. When instructed to smell something, use your hand to waft vapors toward your face and sniff gently.
- Never return reagents to the original container once they have been removed.

SAFE HANDLING OF EQUIPMENT

- Never pipet by mouth. Always use a pipet aid or suction bulb.
- Do not use hot plates with exposed or worn wiring.
- Check Bunsen burner hoses for holes.
- Always ensure balanced loading of test tubes in centrifuges.

ENGINEERING CONTROLS AND PERSONAL PROTECTIVE EQUIPMENT

- Always wear the correct type of eye protection when working in the laboratory. Your instructor will tell you the level of eye protection required.

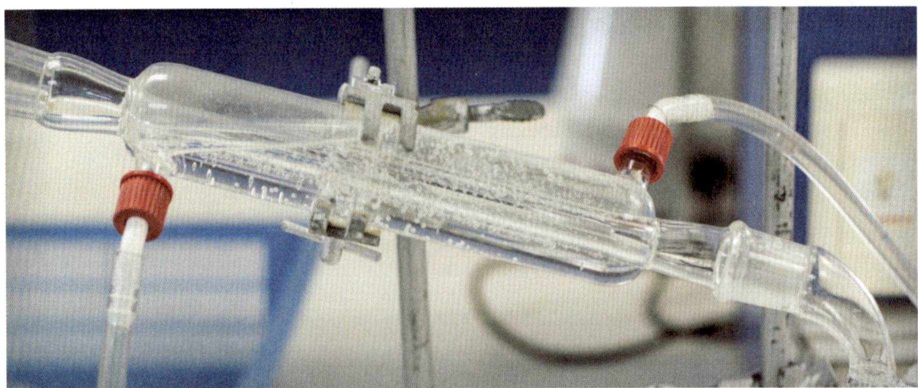

- Wear chemically resistant laboratory coats or aprons, if instructed to do so.
- Work in laboratory hoods as instructed.

PROPER HOUSEKEEPING

- Minimize tripping hazards by keeping aisles free of book bags and other tripping hazards.
- Prevent spills by keeping chemicals and apparatus well away from the edges of your laboratory bench or other workspace.
- Dispose of chemical hazardous waste as instructed, and always ask for guidance if you are unsure.
- Always wash laboratory coats or other clothing on which chemicals have been spilled separately from personal laundry.
- Wipe down your work area for the next user.
- Clean spills on the balances as instructed.

PROPER HYGIENE

- Do not prepare or store (even temporarily) food or beverages in a chemistry laboratory.
- Never consume any food or beverages when you are in a chemistry laboratory.
- Never wear or take laboratory aprons or laboratory coats into areas where food Is consumed.
- Do not chew gum, smoke, or apply cosmetics or lip balm in the laboratory. Be aware that cosmetics, food, and tobacco products in opened packages can absorb chemical vapors.
- Never take your hands or pen to your face or mouth while working in the laboratory.
- Do not handle contact lenses in the laboratory, except to remove them when an emergency requires the use of the eyewash fountain or safety shower.
- Always wash your hands and arms with soap and water before leaving the laboratory, even if you wore gloves.

EMERGENCY PREPAREDNESS

- Become thoroughly acquainted with the location and use of safety equipment and facilities such as exits, evacuation routes, safety showers, eyewash fountains, fire extinguishers, and spill kits.

REFERENCES

[1] Dartmouth Toxic Metals Superfund Research Program. A Tribute to Karen Wetterhahn. www.dartmouth.edu/~toxmetal/about/tribute-to-karen-wetterhahn.html (accessed March 6, 2017).

CHAPTER 3

Guide to Chemical Hazards

Introduction

Chemicals are the tools that chemists use to perform their work. As you read this chapter, you will begin to understand what chemical hazards are. A **hazard** is a potential source of danger or harm. If chemical hazards go unrecognized, unexpected events resulting in personal injury and/or property damage can (and do) occur. **Risk** is a combination of the likelihood of an unwanted incident occurring, the severity of the consequences if it occurs, and the frequency of exposure to the hazard. The fact that a chemical might have an inherent hazard does not mean that we cannot use it in the laboratory. However, an uncontrolled hazard presents increased risks, which may be dangerous. Interestingly, the very properties that make a chemical useful are often those that make it risky to use, so chemists must learn how to safely use chemicals that have significant inherent hazards, by using the principles of RAMP. So, although many chemicals have hazards, most do not present risk in our daily lives with normal use, because we have learned how to recognize hazards and minimize their risks.

If you initiate a conversation with a chemist about chemical hazards, it is likely that you will hear at least one tale about some occasion when that person had to deal with a hazardous situation created by chemical use. All chemicals have inherent hazards that can cause harm if they are not handled properly. Chemists use many methods to minimize or control the risks associated with working with chemicals. To learn how to handle chemicals correctly, you must first be able to identify and understand the hazards present.

An event that is considered the worst industrial incident ever (based on loss of life and suffering) illustrates what can happen when hazards are not properly managed. In December 1984, 40 tons of methyl isocyanate (MIC), a water-reactive chemical, was being stored in a large tank in a non-operational pesticide plant in Bhopal, India. The plant was being decommissioned, and some of the safety features controlling the hazards of the contents of the tank had been disabled. When water

leaked into the tank, a violent reaction occurred and a plume of toxic gases was released into the surrounding community. In the aftermath, it is estimated that 3800 people died immediately, 15,000 died later, and 500,000 were injured. The long-term effects still plague the people of Bhopal today.[1] Water is the universal solvent, the basis for life on Earth, and we would not survive without it. However, under the circumstances that occurred in Bhopal, water mixed with water-reactive MIC, producing a disaster.

The Globally Harmonized System of Classification and Labelling of Chemicals (GHS), implemented in the United States in 2012, is now used to define physical, health, and environmental hazards for each chemical manufactured or sold in the United States. The GHS hazard rating system was developed for a variety of reasons. Foremost among these was to reduce the risk presented by chemicals in the workplace by improving the quality of known information and standardizing the way in which chemical hazards are communicated to workers. ***It is the responsibility of the manufacturer or importer to identify and communicate the hazard(s) of each chemical they produce or sell. It is up to the user (you) to understand the information provided on the label and in the Safety Data Sheet (SDS).***

There are more than 126 million chemical substances registered at the Chemical Abstracts Service (CAS), a division of the American Chemical Society, and this number increases significantly every year.[2] According to the National Toxicology Program, the agency that evaluates health effects for chemical agents of concern, more than 80,000 chemicals are registered for use in the United States, and 2000 more are added each year.[3]

Only a very small fraction of the chemicals in use have been evaluated for their potential to cause harm.[4] Relatively few of the consumer chemicals we are exposed to in everyday use are thought to pose health hazards at these consumer-use levels. However, the hazard may be more significant to a worker in a laboratory who is using the pure form of a partially evaluated chemical if steps are not taken to minimize the risks of exposure.

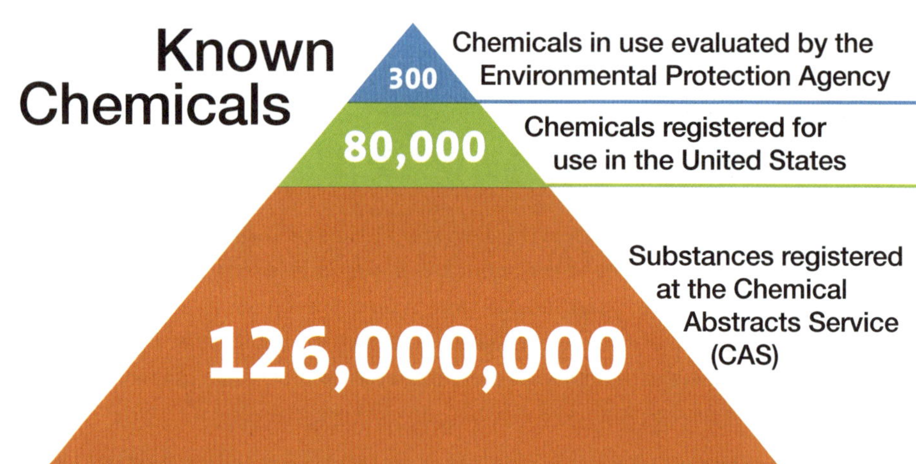

Although not proportional, this figure illustrates that whereas there are millions of chemicals known, we are exposed to very few of them in our daily lives and very few have been fully evaluated for risk to humans. In an analogy, let's say that a 24-foot round swimming pool with 4 feet of water contains 13,600 gallons of water and this represents the 126,000,000 known chemicals. Of this volume, 8 gallons represents the 80,000 chemicals in commercial use. Only about 8 tablespoons or 1/2 cup of that whole pool volume would represent the 300 chemicals adequately tested for safety!

Each newly registered substance is assigned a unique CAS Registry Number, and each substance has its own hazardous characteristics. If you intend to prevent incidents when working with chemicals in the laboratory, then you need to begin to learn about and understand the hazardous characteristics of the chemicals you will work with. In your introductory and organic chemistry laboratories, you will work with several dozen or more chemicals. How can you be expected to know the hazardous characteristics of so many different chemicals? The answer: classification. The hazardous characteristics of all chemicals can be sorted into just a few classes.

Let's look at four broad subclasses of chemical hazard: **toxicity**, **flammability**, **corrosivity**, and **reactivity**. Some chemicals are hazardous in only one of these ways, and some are hazardous in more than one way. Many chemicals used in chemistry laboratories are hazardous in at least one of these ways, but the degree of hazard varies — it can be great, small, or in between. For example, compare gasoline and alcohol with respect to the physical hazard, flammability. Both are flammable liquids, but gasoline is much more hazardous. Gasoline is easier to ignite and more likely to burn vigorously or explode than alcohol, but we safely use gasoline every day. From this, you should understand that we can, and do, know how to safely handle even dangerous chemicals.

Toxicity

Exposure

It has long been known that exposure to any substance in sufficient quantity can be lethal. In the 16th century, a military surgeon and alchemist known as Paracelsus wrote, "What is it that is not poison? All things are poison, and nothing is without poison. It is the dose only that makes a thing not a poison." This means your exposure dose (or amount) to a chemical will determine the toxic effects that you experience.

Although any substance does have the potential to be harmful to those working with it, complex relationships exist between a substance and its physiological effect in each person.

The study of the adverse effects of a substance on living organisms and the ecosystem is known as ***toxicology***.

There are many factors that determine how you, as a living organism, will react when a chemical substance enters your body. Included are such things as how the chemical enters your body (called the route of entry), the amount of substance (the dose) and the length of time for which you are exposed (the duration), the physical state of the toxicant (the form), and many other factors, such as the gender of the exposed person, the stage in the reproductive cycle, age, lifestyle, previous sensitization, which organ is affected, allergic factors, and the individual's genetic disposition – to name just some. These factors all affect the severity of an exposure.

The toxic effects can be immediate or delayed, reversible or irreversible, and local or systemic. The toxic effects vary from mild and reversible (e.g., a headache from a single episode of inhaling ethyl acetate vapor, which disappears when the person inhales fresh air) to serious and irreversible (e.g., birth defects from excessive exposure to nicotine during pregnancy, or cancer from excessive exposure to formaldehyde).

Except for allergic responses, the toxic effects from exposure to a chemical depend on the severity of the exposure (remember Paracelsus). Generally, the larger or more frequent the exposure, the more severe the result. Consequently, you can reduce or even avoid harm by keeping exposures to a minimum.

Now, let's take a close look at some of the above-mentioned factors that can determine how exposure to a chemical substance can adversely affect you.

Routes of Entry/Exposure

The way in which a chemical substance enters the body, called the ***route of entry/exposure*** (ROE), will often determine other factors of exposure. In addition, it is important to know how a chemical might be

introduced into your body in order to protect yourself from exposure, because your risks can be reduced by eliminating or minimizing each of the routes of exposure.

There are four ways in which chemicals can enter the body:

Inhalation. A chemical enters the body through the respiratory tract, by breathing. The substance can be in the form of a vapor, gas, fume, mist, or dust. This is considered the most common ROE in chemistry laboratories.

Ingestion. A chemical enters the digestive tract through the mouth (orally). It is unlikely that one would ingest a chemical in the laboratory on purpose, and there are basic rules to prevent accidental ingestion of chemicals while in the laboratory. Exposure to chemicals via this route can occur through eating, chewing gum, applying cosmetics, or smoking in the laboratory (which is not as big of a problem as it once was), or eating lunch without washing your hands after working in the laboratory. This ROE is eliminated by prohibiting eating and drinking in the laboratory.

Absorption. When a chemical comes in contact with the skin, dermal absorption of the chemical may occur. Absorption of chemical vapors can also occur through the eyes and mucous membranes.

Injection. Chemicals enter the body through a cut made in the skin by a sharp contaminated object. Possibilities include mishandling a sharp-edged piece of a contaminated broken glass beaker or misuse of a sharp object, such as a knife or hypodermic needle.

Dose

For chemicals, **dose** is defined as the amount of toxicant received at one time. Dose is commonly reported in terms of amount per body mass, such as milligrams per kilogram (mg/kg); it is normalized on body mass so that it can be compared with other dose reports. But dose may be reported in other ways for other routes of exposure. For example, skin or dermal doses are usually reported in terms of amount per skin surface area, such as milligrams per square centimeter (mg/cm^2). Airborne doses are usually reported in terms of amount per unit volume of air (concentration), such as micrograms per liter ($\mu g/L$), milligrams per cubic meter (mg/m^3), or parts per million (ppm) for a given time period.

Duration and Frequency of Exposure

The health effects from a toxicant can be described by the duration of exposure and the onset of the effect.

Acute exposure is characterized by rapid assimilation of the toxic substance in one or more doses within 24 hours or less. Typically, the resulting effect has a sudden onset and is localized, and it can be painful, severe, or even fatal. Usually, a single exposure to a high concentration is involved (see the definition of "lethal dose" in "In Your Future: Toxicological and Regulatory Terms" at the end of this chapter).

The effects of an acute exposure are often reversible. For example, if you inhale a toxicant and you immediately experience difficulty breathing but your breathing returns to normal when you leave the room and get fresh air, this would be described as an acute exposure resulting in an acute effect. However, if the toxicant gets into your bloodstream and results in a ***systemic effect*** in another organ, the effect may not manifest immediately. In this case, an acute exposure might have a ***delayed effect*** or a ***chronic (long-term) effect***.

Chronic exposure is characterized by repeated exposures, typically of low doses, with a duration measured in months or years. The effects of the exposure may not be immediately apparent (said to be *insidious*) and are typically not reversible.

Pharmacokinetics is the study of how the body processes substances to which it is exposed. Once the substance is in the body, it will go through a defined process: absorption, distribution, metabolism, and excretion (called ADME). How rapidly and where absorption takes place, what organs the substance is distributed to, how it is (or is not) metabolized (converted into other substances), what metabolites are formed, and how quickly it can be excreted will all affect how toxic the substance will be for that individual.

Groups of Chemicals Known to Elicit Toxic Effects

A ***synergistic effect*** occurs when two (or more) chemicals combined produce an adverse effect that is greater than that expected if you were to add together the effects of the individual chemicals. An example of synergy is exposure to alcohol and chlorinated solvents: the alcohol increases the toxicity of the chlorinated solvent. The opposite is also possible: one toxic substance can lessen another's effect, in an ***antagonistic effect***. There are several mechanisms of antagonism. One example of an antagonistic effect is the use of ethanol as an antidote for methanol ingestion. The metabolites of methanol are toxic, but because ethanol is preferentially metabolized, the methanol can be excreted.

Allergens are agents that produce an immunological reaction, and you may encounter them in the laboratory. An allergen can cause a respiratory asthma-like response or a contact dermatitis (eczema) reaction. Not everyone is susceptible to allergens. A chemical is said to be a ***sensitizer*** if it elicits an allergic response in a significant population. Reactions to poison ivy are allergic responses. Common sensitizing chemicals to which you might be exposed in a chemistry laboratory are nickel metal, sulfur and its compounds, salicylates (aspirin and wintergreen), formalin (formaldehyde), and latex (which is used less frequently now). Tell your instructor if you know or suspect that you might be allergic to a chemical that will be used in your laboratory — yet another reason why you should read the experimental procedure before coming to the laboratory.

Lachrymators are chemical substances that cause prolific tearing of the eyes due to their profound effect on the lachrymal glands. You are likely familiar with one of these substances (1-sulfinylpropane) if you have ever cut onions. Tearing is a biological response that attempts to dilute the irritating substance. In all but the most severe exposures, the effects of lachrymators do not result in permanent damage to the eye. If a chemical is a lachrymator, this information should be indicated on the label and in the SDS. Goggles will not necessarily prevent the vapors from initiating the response, and you should work with these substances only in a laboratory hood. Glassware should be rinsed in the hood before removing it to be washed in the sink. You should always wash your hands at the end of laboratory work, but it is especially important when you have been working with these chemicals, to ensure that these chemicals do not come in contact with the eyes.

Organic solvents can penetrate intact skin and are easily inhaled; they therefore present a health hazard in addition to a flammability hazard. Many organic solvents can penetrate intact skin and are easily inhaled when sufficiently volatile; they therefore present a health hazard in addition to a flammability hazard posed by many of these solvents. When in contact with the skin, most organic solvents cause dryness and cracking. The vapors of all organic solvents are toxic, some more than others. Typical signs and symptoms of overexposure to organic solvent vapors include headaches, dizziness, slurred speech, changes in breathing or heart rate, unconsciousness, and, rarely, death. Typical target organs affected by organic solvents are the central nervous system, the liver, and the kidneys. Avoid skin contact with these liquids. Work with organic solvents should be carried out in a laboratory hood to keep vapors in the breathing air at acceptable levels. Gloves must be chosen carefully to make sure they are adequately protective.

Heavy metals have numerous known toxicological effects. You might still encounter *elemental mercury* from a broken thermometer (if your laboratory still uses mercury thermometers) or a spill from a manometer, used to measure pressure in a chemistry laboratory. More and more, academic laboratories are replacing mercury-containing devices with safer alternatives because of the hazard. Mercury is a cumulative neurological toxicant. Exposures can be caused by absorption through the skin and inhalation of the vapor. When spilled, mercury forms hard-to-contain droplets, some of which are too tiny to be seen. Spilled mercury must be immediately and thoroughly cleaned up by properly trained individuals using specialized equipment and detection methods. Notify your instructor immediately if you break a mercury thermometer or see a broken thermometer.

Asphyxiants are substances that have the ability to deprive the body of oxygen. A simple asphyxiant (such as nitrogen) displaces or dilutes oxygen in air to a level not compatible with life. A chemical asphyxiant (such as carbon monoxide) either prevents the body from using the oxygen available in air or impairs oxygen transport in the body.

Flammability

Solvents

Solvents are liquids that are used to dissolve or disperse other reagents. Organic solvents represent a large class of liquids that you will work with, especially in organic chemistry laboratories. Many organic solvents present significant flammability hazards. Flammable solvents such as acetone, hexane, methanol, ethanol, and acetonitrile are commonly used in chemical teaching and research laboratories.

Organic solvents can be divided into three broad types: those that contain only hydrogen and carbon (hydrocarbons), those that contain oxygen as well (oxygenated solvents), and those that contain halogens (halogenated solvents). Many (but not all) of the halogenated solvents (e.g., methylene chloride, carbon tetrachloride, and chloroform) are not flammable but are quite toxic. Hydrocarbons (e.g., hexane, toluene, and xylene) and oxygenated organic solvents (e.g., methanol, diethyl ether, and acetone) are typically very flammable. ***Do not*** rely on generalizations about flammability; always check the label on a solvent you are using, because it will indicate the flammability.

It is very important to understand that a flammable liquid itself cannot burn; it is the vapor (the gaseous form of the chemical) from the liquid that burns. The rate at which a liquid produces flammable vapors depends on its rate of vaporization, which increases as the temperature increases. Consequently, a flammable liquid is more hazardous at elevated temperatures than at normal temperatures. Many organic solvent vapors are denser than air and can travel to a source of ignition and "flash back". So, remember that when you are pouring out a flammable solvent, you are also pouring out the invisible, flammable vapors, which, if exposed to a nearby source of ignition, can ignite into a flash fire.

All flammable liquids and solids must be kept away from oxidizers and from inadvertent contact with ignition sources, such as hot plates in hoods. Do not store stock containers of solvent in the hood where you are working. Flammable organic solvents should be stored at room temperature in a flammable cabinet unless other storage conditions are indicated on the manufacturer's label.

Flammable Solids

Unlike pyrophoric solids (discussed in the "Reactivity" section), flammable solids require an ignition source. Many metals commonly used in teaching laboratories, such as iron, magnesium, calcium, and aluminum, are flammable. The more finely divided the material is, the greater the risk. Flammable metal fires are easy to initiate and difficult to extinguish, requiring specialized extinguishing materials. Never place finely divided metals into trash cans with combustible materials; in fact, you should not be placing any chemicals into trash cans.

Corrosivity

Corrosives

Corrosion is the gradual destruction resulting from the action of a chemical on metal or living tissue. All strong acids (e.g., hydrochloric acid, sulfuric acid, and nitric acid), all strong bases (e.g., sodium hydroxide and potassium hydroxide), some weak acids (e.g., acetic acid, carbonic acid, and phosphoric acid), some weak bases (e.g., ammonium hydroxide), and some slightly soluble bases (e.g., calcium hydroxide) are corrosive.

Even an acute exposure to a corrosive chemical can irreversibly destroy living tissue. Your eyes are particularly vulnerable. The more concentrated the acid or base and/or the longer the contact, the greater the destruction. Some acids and bases initiate damage within 15 seconds of contact. For this reason, you should always wear chemical splash goggles when handling corrosives. Chemistry laboratories where corrosive substances are used are required to have an eyewash fountain and safety shower unit accessible from any point within a 10-second time frame. You should learn where the eyewash/shower is and practice walking there, maybe with your eyes closed.

The corrosive solutions that you are most likely to encounter in various concentrations in your first chemistry laboratories are hydrochloric acid, sulfuric acid, phosphoric acid, acetic acid, nitric acid, sodium hydroxide, and ammonium hydroxide.

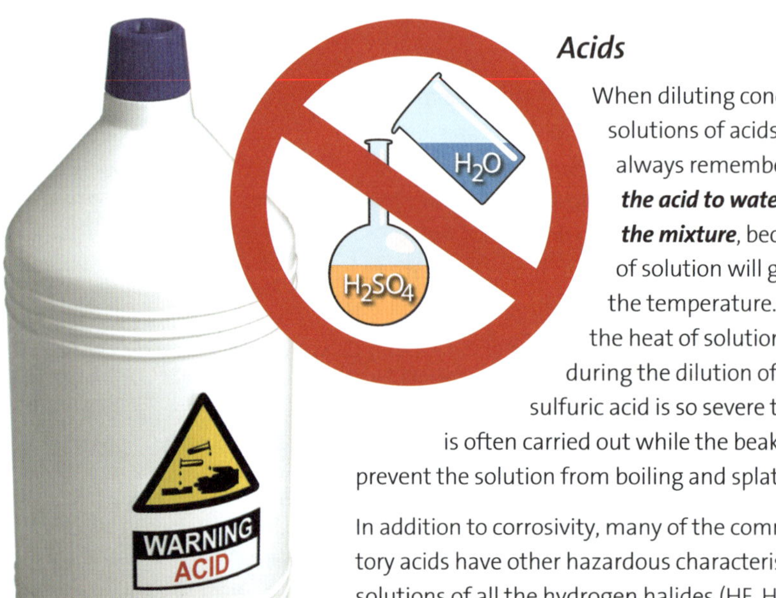

Acids

When diluting concentrated solutions of acids, you must always remember to **slowly add the acid to water while stirring the mixture**, because the heat of solution will greatly increase the temperature. For example, the heat of solution evolved during the dilution of concentrated sulfuric acid is so severe that the process is often carried out while the beaker is on ice, to prevent the solution from boiling and splattering.

In addition to corrosivity, many of the common laboratory acids have other hazardous characteristics. Aqueous solutions of all the hydrogen halides (HF, HCl, HBr, and HI) are toxic, but HF is of special concern (see "In Your Future: Acids That Can Be Particularly Dangerous"). The vapors of these acids are serious respiratory irritants.

Concentrated sulfuric acid is a very strong dehydrating agent (able to remove water), and all except very dilute solutions can be oxidizing (see the "Reactivity" section). Phosphoric acid is a weak acid. The concentrated acid is a viscous liquid and, like sulfuric acid, is a strong dehydrating agent.

Nitric acid is also a strong oxidizing agent. It generally reacts more rapidly than sulfuric acid does. If dilute nitric acid gets on the skin and is not washed off completely, it causes the exposed skin to become yellowish brown as a protein-denaturing reaction occurs. Nitric acid is discussed further in the "Reactivity" section.

■ IN YOUR FUTURE: Acids That Can Be Particularly Dangerous

It is unlikely that you will encounter perchloric acid, picric acid, or hydrofluoric acid (HF) in your first chemistry laboratories. Each of these acids has multiple severe hazards. Their use in academic research laboratories is not as common as it once was, but when they are needed, their use and storage must be carefully controlled. The harmful consequences of improper handling of these reagents are great. Perchloric acid and picric acid are also discussed in the "Reactivity" section.

HF is toxic and is rapidly absorbed through the skin, where it penetrates deeply and destroys the underlying tissues. Contact with a dilute solution of HF is usually painless for several hours, followed by serious burns, adverse internal effects (including bone destruction), and excruciating pain. Exposures to this acid to 25% of the body can result in death. You should never handle HF without having a full understanding of the hazards, very specific training, and proper personal protective equipment (PPE), and ensuring that emergency response procedures are in place.

Bases

The most common bases used in academic laboratories are the alkali metal hydroxides and aqueous solutions of ammonia. Sodium hydroxide and potassium hydroxide are strong alkali bases and are extremely destructive to the skin and the eyes. Ammonia in aqueous solution, commonly referred to as ammonium hydroxide, is a weak base. The vapors of aqueous solutions of ammonia are irritating and toxic. Aqueous solutions of ammonia are particularly damaging to the eyes.

Strong bases are all corrosive and can cause serious, destructive chemical burns, including blindness if splashed into the eyes. Bases do have good warning properties: they typically have a slippery feeling, because of the saponification of the oils in your skin, and hence you know to keep rinsing until that feeling is gone. However, if it is not completely removed by rinsing, a solution of a strong base may not cause pain until the corrosive damage is quite severe.

Reactivity

Chemicals have the ability to react with other chemicals and transform into new substances. This is the basis of all chemical experimentation. Reactivity in and of itself is not necessarily a concern, but ***uncontrolled reactivity*** is a big concern. To manage reactivity, you must learn to recognize certain properties of chemicals. Not all reactivity is based on chemicals reacting with each other. Some chemicals are self-reactive, and others are unstable and decompose vigorously if disturbed. Reactivity includes all these characteristics. As mentioned in the introduction to this chapter, even nonhazardous chemicals (such as water) can present a hazard given the right circumstances.

As you are learning in this chapter, chemicals have hazardous properties that must be managed in use and in storage. Most experiments in chemistry laboratories use chemicals that react with each other, and these reactions are useful and, sometimes, hazardous. Acids react with bases; oxidizing agents react with reducing agents. When these reactions present particular hazards, we often refer to the two reactants as "incompatible". Depending on the degree of incompatibility, the reaction may be mild or very vigorous. Many of these incompatibilities have been documented in the literature and on the Internet. Internet resources can easily be found by simply searching for "chemical incompatibility chart". A list of additional published resources is given in the Appendix.

The concept of incompatibility also plays an important role in how chemicals are stored in a research laboratory or stockroom. One's first inclination is to store chemicals alphabetically or by experiment, but this can result in storing incompatible chemicals next to each other. This presents a potentially hazardous situation if containers leak or break and the chemicals mix and react. Chemical storage systems segregate categories of chemicals to keep incompatible chemicals well separated from each other. Again, numerous resources on this subject are available in print and online

Oxidizers

One large class of reactive chemicals that you are likely to encounter, even in your first chemistry laboratories, are those classified as oxidizers. These substances either can supply oxygen in a reaction or can be reduced (gain electrons), thereby facilitating oxidation (loss of electrons) of another substance. As you learn about chemistry, you will be required to learn nomenclature – the language of chemistry. Many chemicals in this class have names that end in "ate", "ite", or "ic", or begin with "per". This is only a broad characterization and should not be taken as a certainty.

Some common oxidizing solutions you might encounter in your first experiments are various metal salts of nitrates, potassium permanganate, hydrogen peroxide, bromine (in the organic chemistry laboratory), sulfuric acid, and nitric acid. Of particular concern with oxidizing agents is proper storage. They should always be stored separately from any chemical or material that can be oxidized. This includes not storing bottles on wooden shelves or with flammable solvents.

Peroxide-Forming Solvents

A peroxide is a compound that contains two oxygen atoms joined with a single bond (–O–O–). You may be familiar with the most common of these compounds: hydrogen peroxide. In organic peroxides, one or both of the hydrogen atoms has been replaced with an organic group. All peroxides are reactive, but the organic peroxides are particularly so because they pose unusual stability problems. It is unlikely that you will encounter this hazardous class of chemical in your first experiments, but **you might work with a solvent that can form unstable peroxides over time when exposed to air or light.**

A few organic solvents (e.g., ethers and some non-aromatic unsaturated cyclic hydrocarbons) can form potentially explosive hydroperoxides and peroxides. These solvents are particularly dangerous if they are evaporated close to dryness. You may use diethyl ether in your organic chemistry laboratory, and your teaching laboratory manager should ensure that the solvent is peroxide-free. However, if you have any doubts, ask your instructor.

In any case, until you know which organic solvents form peroxides, do not move or open old containers of organic solvents if you encounter them. All organic peroxides are extremely flammable, and some are shock-sensitive and explosive. A peroxide present as a contaminating reagent in a solvent can change the course of a planned reaction. Alert your instructor or advisor immediately if you see crystals around the cap or crystals in an organic solvent.

■ IN YOUR FUTURE: Particularly Dangerous Hazards

As mentioned in the "Corrosivity" section, some less commonly used acids, such as those listed here, are also reactivity hazards.

Perchloric acid *is a very powerful oxidizing agent at elevated temperatures, and it can react explosively with organic compounds and other reducing agents. Use of this acid for sample digestion requires special facilities and training.*

Picric acid *is a solid acid with a chemical structure similar to that of trinitrotoluene (TNT). This acid must always remain wetted, because dry picric acid is explosive. If you encounter a bottle of this hazardous solid and cannot determine the state or age, immediately seek assistance from your instructor or advisor. This is not a chemical that should be handled by an untrained student.*

Pyrophoric materials *are very reactive reagents that are hazardous, even in small quantities, because they are liable to ignite very shortly (within 5 minutes or less) after contact with air. These compounds require specialized training before use. Examples of these compounds are organolithium compounds (e.g., tert-butyllithium solution) and some very fine metal powders (e.g., magnesium).* ***These substances should only be used with documented advanced training and advisor oversight.*** *These compounds will have the GHS hazard code H250 (see "Globally Harmonized System of Classification and Labelling of Chemicals (GHS)").*

Shock- and friction-sensitive materials *as a separate class have not been specifically discussed, but some were mentioned in the "Reactivity" section. Some well-known chemicals in this class (broadly stated) are metal azides, perchlorate salts, and organic peroxides. Use of these materials should be controlled as stated above.*

Recognizing Chemical Hazards: Sources of Information

Your Instructor or Advisor

The instructor or advisor in charge of each laboratory is a very important resource for chemical safety information. As the initial point of contact in the laboratory, an instructor is prepared to explain the hazards associated with the laboratory chemicals in use and to give you the necessary precautions to take that will reduce risk and prevent exposure.

Globally Harmonized System of Classification and Labelling of Chemicals (GHS)

The GHS[5] is all about communicating hazards to users; remember the R and the A of RAMP: Recognize hazards, and Assess the risks of hazards. Using the GHS for hazard recognition requires that you have a basic understanding of the elements of the system. In the GHS, there are 17 physical hazard classes, 10 health hazard classes, and 2 environmental hazards classes. Within each class, the hazard is placed into a category based on various criteria specific to that classification. Each category is assigned a number or a letter, for example 1 to 5 or A to E. In the GHS, ***the lower the category value is within each classification for a chemical, the more severe the hazard***.

The hazard categories are communicated to the user through pictograms, hazard statements, precautionary statements, and signal words. Hazard categories are especially helpful in assessing the relative risks of hazards. For example, acetonitrile is classified under the GHS as a flammable liquid (category 2) with the flame pictogram and the signal word "Danger". This solvent is also rated as a category 4 acutely toxic chemical by the oral ROE. Remember that category 1 or A is the most hazardous rating for that class.

The U.S. Occupational Safety and Health Administration (OSHA) defines a chemical presenting a **physical hazard** as "a chemical that is classified as posing one of the following hazardous effects: explosive; flammable (gases, aerosols, liquids, or solids); oxidizer (liquid, solid, or gas); self-reactive; pyrophoric (gas, liquid, or solid); self-heating; organic peroxide; corrosive to metal; gas under pressure; in contact with water emits flammable gas; or combustible dust."

OSHA defines a chemical presenting a **health hazard** as "a chemical that is classified as posing one of the following hazardous effects: acute toxicity (any route of exposure); skin corrosion or irritation; serious eye damage or eye irritation; respiratory or skin sensitization; germ cell mutagenicity; carcinogenicity; reproductive toxicity; specific target organ toxicity (single or repeated exposure); or aspiration hazard."

GHS Pictograms and Hazards

Health Hazard	**Flame**	**Exclamation Mark**
◆ Carcinogen ◆ Mutagenicity ◆ Respiratory Sensitizer ◆ Reproductive Toxicity ◆ Target Organ Toxicity ◆ Aspiration Toxicity	◆ Flammables ◆ Pyrophorics ◆ Self-Heating ◆ Emits Flammable Gas ◆ Self-Reactives ◆ Organic Peroxides ◆ Desensitized Explosives	◆ Irritant (skin and eye) ◆ Skin Sensitizer ◆ Acute Toxicity (harmful) ◆ Narcotic Effects ◆ Respiratory Tract Irritant ◆ Hazardous to Ozone Layer (non-mandatory)

Gas Cylinder	**Corrosion**	**Explosive Bomb**
◆ Gases Under Pressure	◆ Skin Corrosion/Burns ◆ Eye Damage ◆ Corrosive to Metals	◆ Explosives ◆ Self-Reactives ◆ Organic Peroxides

Flame Over Circle	**Skull and Crossbones**	**Environment (non-mandatory)**
◆ Oxidizing Gases, Liquids, and Solids	◆ Acute Toxicity (fatal or toxic)	◆ Aquatic Toxicity

Elements of the GHS

Pictograms are pictures that represent a concept. The GHS uses nine pictograms to visually alert users to the chemical's hazard class. GHS pictograms, along with the hazard classes they cover, are shown in Figure 1. The degree of each hazard for each chemical within each class must be evaluated by the manufacturer. If the degree of hazard is great enough within a class, then that pictogram is required on the label and in the SDS. There is no minimum or maximum number of pictograms that a substance may warrant.

Hazard statements are short statements that describe each physical, health, and/or environmental hazard. There are quite a few hazard statements, and each one is assigned an H code as an alphanumeric identifier.

Precautionary statements are short statements that indicate how to handle, store, prevent exposure to, and dispose of a substance. There are even more precautionary statements than hazard statements. Each of these is assigned a P code as an alphanumeric identifier.

Signal words provide the user with an immediate indication of the hazard severity in each class. There are two signal words: "Danger" and "Warning". Within a specific hazard class, "Danger" is used for the more severe hazards, and "Warning" is used for the less severe hazards.

■ **IN YOUR FUTURE: Understanding the Globally Harmonized System**

With so many hazard and precautionary statements, how can you know them all? The short answer to this question is that you do not need to memorize the codes, because the text of the hazard statement or precautionary statement is required on information that accompanies a hazardous chemical. However, if you do happen to see an H code or a P code by itself somewhere, there is a quick way to at least identify the basic classification represented.

All hazard codes begin with the letter H, followed by three digits. The first digit after the H indicates whether the hazard is a physical (2), health (3), or environmental (4) hazard. You can also learn to recognize groupings within a class. For example, codes 220 to 230 indicate some type of flammability hazard.

All precautionary statements begin with the letter P, followed by three digits. Here, the first digit after the P indicates something general (1), for prevention (2), for response (3), for storage (4), or for disposal (5).

Manufacturer's Container Labels

The last thing that stands between hazard recognition and the user is right at your fingertips: the label on the container that the chemical is stored in. This is why OSHA's Hazard Communication Standard (29 CFR 1910.1200, Appendix C) was aligned with the GHS and required the labels placed on hazardous chemicals to conform to a standard format. As of June 1, 2015, hazardous chemicals are required to have the specific *label elements* required by this standard.

The following items based on the elements of the GHS must be present on the manufacturer's label:

- The **product identifier**. This is typically the International Union of Pure and Applied Chemistry (IUPAC) name, but a trade name or common name for the chemical may be given instead. The CAS Registry Number is almost always included as well. The identifier on the label ***must exactly match the one in the SDS (Section 1).***

- The **supplier's information**. This includes the name, address, and telephone number of the manufacturer, importer, or responsible party.

- **One signal word**, if required, as determined by the level of hazard in any class. Although a chemical might have multiple hazards that warrant a signal word, ***only the greatest level of hazard will be represented on the label***.

- Each **hazard statement**, as text. This is required on the label except as otherwise specified in the regulation.

- All **pictograms**, as determined by the level of each hazard present.

- Each **precautionary statement**, as text.

A label may also include **supplemental hazard information**, as determined by the manufacturer.

Safety Data Sheets (SDSs)

Under the GHS, OSHA defines a **hazardous chemical** as "any chemical which is classified as a physical hazard or a health hazard, a simple asphyxiant, combustible dust, pyrophoric gas, or hazard not otherwise classified."

The SDS for a hazardous chemical is a document that describes the chemical's hazards and the precautions that you must take to avoid harm. OSHA requires employers to maintain an SDS for each hazardous chemical on the premises, available to any employee who requests it. As a student, you can also request the SDS for a chemical. The Internet also makes it very easy to search for and find an SDS online, and they can be very educational.

A GHS SDS is divided into 16 sections. The order of the information presented in an SDS is mandated by regulation, so the information given is relatively uniform from manufacturer to manufacturer. The SDS must be in English. A summary of the 16 sections that must be in an SDS is shown in "The Design of a Safety Data Sheet".

■ The Design of a Safety Data Sheet

Below is a summary of the information that must be provided by the manufacturer in the Safety Data Sheet. OSHA has prepared a Brief that provides a detailed description of the contents of each section.[6]

Section 1: *Identification*

Section 2: *Hazard(s) Identification*

Section 3: *Composition/Information on Ingredients*

Section 4: *First-Aid Measures*

Section 5: *Fire-Fighting Measures*

Section 6: *Accidental Release Measures*

Section 7: *Handling and Storage*

Section 8: *Exposure Controls/Personal Protection*

Section 9: *Physical and Chemical Properties*

Section 10: *Stability and Reactivity*

Section 11: *Toxicological Information*

Section 12: *Ecological Information*

Section 13: *Disposal Considerations*

Section 14: *Transport Information*

Section 15: *Regulatory information*

Section 16: *Other Information*

*Exposure guidelines are established for many chemicals and will be included in an SDS if they are known. Guidelines can be based on basic health hazard information, legal or recommended limits of exposure set by various agencies, or animal studies. PELs, TLVs, TWAs, STELs, and C values (explained below) are for air contaminants. When evaluating the numerical values, do not consider them as some magical cutoff line between a "safe" exposure and an "unsafe" one. In addition, the air must be sampled to determine these values in your breathing air. For most chemicals, this is not a trivial measurement. Often in an SDS you will see the statement "Data not available" in the Toxicological Information section. **Do not** assume that this equates with "safe"; it simply means that the substance has not been evaluated with respect to that value.*

__Lethal dose__ refers to the measurement of an adverse effect in a test population of a specific animal species. Lethal dose, 50% (LD_{50}) is the calculated single dose (milligrams of substance per kilogram of body mass) expected to result in the mortality of 50% of the test population when administered by any ROE other than inhalation. For inhalation, the dose is the lethal concentration, 50% (LC_{50}). It is the concentration of a chemical in breathing air calculated to result in the mortality of 50% of the test population exposed over a specific time period.

__Permissible exposure limits (PELs)__ are established by OSHA, and they are considered to be the maximum concentration of a toxicant that can be inhaled without harm by an adult worker for 8 hours a day, 40 hours a week, during his or her working lifetime. The PEL is a legal limit.

__Threshold limit values (TLVs)__ are established by the American Conference of Governmental Industrial Hygienists (ACGIH). The TLV is a voluntary, recommended limit, and TLVs may differ from PELs. Most authorities have noted that __TLVs are more reliable for protection than PELs__ because they are revised annually, whereas the PEL list is rarely revised.

The __time-weighted average (TWA)__ is a practical average value of worker exposures measured and averaged over an 8-hour workday.

The __short-term exposure limit (STEL)__ is the concentration in parts per million (ppm) or milligrams per cubic meter (mg/m^3) that should not be exceeded for more than a short period (usually 15 minutes) during the 8-hour workday.

__Ceiling limits (C values)__ are assigned to some chemicals with very severe health hazards. The ceiling limit is a value that must not be exceeded at any time, regardless of duration.

SUMMARY

All chemicals have inherent hazardous properties, and OSHA has aligned U.S. regulations to use the GHS to classify chemical hazards into three broad categories: physical, health, and environmental hazards. Within the broad classifications of hazards, there are several major subclasses: toxicity, flammability, corrosivity, and reactivity. The risk involved in using chemicals can vary from very low to very high but can be managed if hazards are identified and controls are put in place.

Typically, introductory and organic chemistry experiments have been designed to manage known hazards. With guidance from your instructor and by heeding the precautions that are described on labels, you will be able to work safely in these laboratories. As you progress in chemistry, you will need to learn how to manage greater hazards by developing a deeper understanding of the properties and toxicology of hazardous chemicals by evaluating Safety Data Sheets (SDSs) and using hazard analysis and risk management tools.

When working with chemicals in the laboratory, it is always prudent to prevent or minimize chemical exposure. This is done by understanding the hazardous properties of the chemicals you are working with, adhering to established best practices, reading safety precautions in experimental procedures, wearing your personal protective equipment (PPE) as instructed (see Chapter 2), and working with volatile organic solvents in a hood, which prevents inhalation – the most common route of entry (ROE) of toxicants into the body.

Broadly speaking, there are four classes of exposure–effect relationships:

- an acute exposure resulting in an immediate acute effect (e.g., cyanide poisoning);
- an acute exposure resulting in a delayed acute effect (e.g., acetaminophen overdose, which is initially asymptomatic, followed by liver failure starting 24–48 hours after the acute ingestion);
- an acute exposure resulting in a chronic effect (e.g., acute exposures to some neurotoxic pesticides can contribute to chronic neurodegenerative disorders, such as Parkinson's disease);
- a chronic exposure resulting in a chronic effect (e.g., alcoholic cirrhosis).

You should realize that, as a new chemistry student, there are limitations to your knowledge about chemical hazards. You are currently at the stage of learning a subject where you don't know what you don't know. To protect yourself and others, ask questions if you are unsure, and read all the safety information you are given, so that you can begin to learn to recognize hazards associated with the chemicals you use.

Finally, you should think of chemicals as useful tools for chemists, and for all of society. In fact, chemicals save lives and improve the quality of life for millions of people each day. We need and use chemicals in much of what we do every day, and we can and must learn to minimize the risks of chemical hazards, to maintain our safety and the safety of others.

REFERENCES

[1] Broughton, E. The Bhopal Disaster and Its Aftermath: A Review. *Environ. Health*. 2005, *4*, 6. http://dx.doi.org/10.1186/1476-069X-4-6

[2] Chemical Substances – CAS REGISTRY. A Division of the American Chemical Society. www.cas.org/content/chemical-substances (accessed March 6, 2017).

[3] About NTP, National Toxicology Program, U.S. Department of Health and Human Services. http://ntp.niehs.nih.gov/about/index.html (accessed March 6, 2017).

[4] Fischetti M. The Great Chemical Unknown: A Graphical View of Limited Lab Testing. *Scientific American*, Nov 1, 2010. www.scientificamerican.com/article/the-great-chemical-unknown (accessed March 6, 2017).

[5] United Nations. Globally Harmonized System of Classification and Labelling of Chemicals (GHS), Fifth revised edition, ST/SG/AC.10/30/Rev.5; New York and Geneva, 2013.

[6] OSHA. Hazard Communication Standard: Safety Data Sheets, 2012. www.osha.gov/Publications/OSHA3514.html (accessed March 6, 2017).

CHAPTER 4

Recommended Laboratory Techniques

Introduction

Chapter 3 described some types of physical, health, and environmental chemical hazards and the effects of being exposed to chemicals. The focus of this chapter is on how to safely perform common laboratory techniques and safely handle the most common equipment in the undergraduate chemistry laboratory. The techniques and advice in this booklet focus on those topics most commonly encountered in first- and second-year courses in college. There is brief mention of some advanced topics that you may encounter in upper-level courses in the "In Your Future" sidebars, but a thorough presentation of advanced techniques is beyond the scope of this publication. The references in the Appendix point to other sources of safety information.

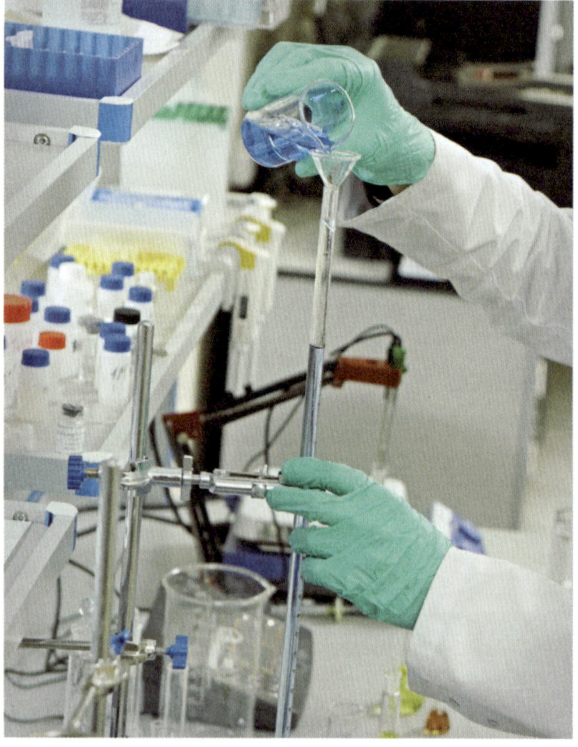

As discussed in Chapter 1, the RAMP system is a useful paradigm when you work in the laboratory. Chemical safety must be the priority of everyone. Apply the RAMP concept as you prepare for each laboratory session. Once hazards have been recognized and assessed, they must be minimized or managed. Minimizing hazards to reduce risk involves adding controls or placing barriers between the worker and the hazard. A control measure is simply the steps taken to eliminate or minimize a hazard. (Note that fewer hazards or less hazardous substances used in the laboratory can also have a positive effect on the environment. See "Green Chemistry — Not Just a Catchphrase".) There is an order, or hierarchy, of controls, used to prioritize the steps that should be considered (see "Hierarchy of Controls").

As an undergraduate, you must carefully listen to all directions and complete all laboratory preparations before coming to the laboratory. Carefully review the details of an upcoming laboratory procedure, and ask your instructor before beginning the experiment if you have any concerns about laboratory hazards. Your instructor or undergraduate research mentor may want you to review *advanced* techniques for a particular experiment and may provide specific instructions for you to follow to ensure your safety.

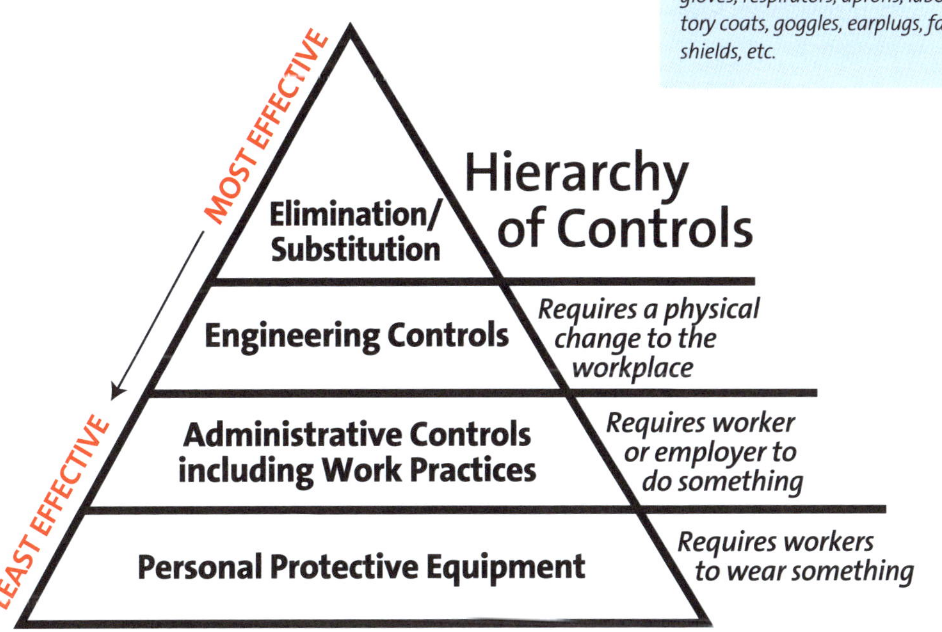

Working with Chemicals, Apparatus, and Equipment

Following these recommendations will help to make your work easier and laboratory equipment safer to use, thereby increasing the likelihood that your experiment will be successful. As described in Chapter 3, there are many hazards potentially present from the chemicals in a laboratory, but the equipment used in experiments can also present hazards. Be particularly alert to procedures where temperature and pressure are elevated or lowered, electrical equipment is required, open flames are used, or you are unfamiliar with the equipment being used. Never disable safety features or guards on equipment.

General Concerns when Working in the Laboratory

As you read over the laboratory directions, look up any new terms and review reference sources suggested by your laboratory instructor. Be sure that you know what to do if you or another student experiences an unexpected occurrence (a sudden rise or drop in temperature, excess gas evolution, etc.). Or, for example, if the instructions for an experiment state that the temperature should *not* exceed a predetermined point and you don't understand why, you should discuss this with your instructor.

Set up your experimental apparatus away from the edge of the laboratory bench, where it could easily be knocked onto the floor. Secure laboratory apparatus when necessary. For example, an Erlenmeyer flask can be placed directly on the laboratory bench, but a round-bottom flask needs to be held with an appropriate clamp attached to a ring stand or placed in a cork ring.

Inadvertent Exposure to Chemicals

Chemicals can get transferred to you indirectly (and unexpectedly) in a wide variety of ways. Some common examples are:

- Chemicals on your hand, or gloved hand, can easily get transferred to a pen or pencil in the laboratory and then get transferred to your mouth if you are in the habit of placing a pen or pencil in your mouth. Avoid this habit.

- Chemicals on your hand can get transferred to your eye if you momentarily raise your goggles to rub your eye. Chemicals on your hands will easily get transferred to a cell phone or computer. Further, the cell phone might then transfer the chemical to your face if it is used as a phone in the laboratory.

- If you wear gloves in a laboratory and they become contaminated, they will spread this contamination to everything that you touch in the laboratory: computers, balances, hot plate controls, glassware, and cell phones. This easily spreads the contamination. Gloves should be removed whenever you leave the laboratory, because otherwise they might contaminate doorknobs and other items outside the laboratory.

The purpose of gloves is to act as an extra barrier to avoid skin contact. Ideally, they should not become contaminated, but if you suspect this, you should replace them immediately to avoid the scenarios listed above.

Scientific Glassware

The directions provided by your instructor should indicate the appropriate size of glassware to accommodate the operation to be performed. If the exact size (e.g., 250 mL) is not available, keep in mind that you should select glassware so that at least 20% free volume remains in the vessel throughout the experiment. For chemical reactions, borosilicate glassware (e.g., Pyrex or Kimax) is used because it can withstand extreme changes in temperatures and minimal shock without breaking.

Clean glassware before use (see Chapter 2). If you share equipment with another student, be particularly aware of the condition that your glassware is in each period. You should examine your glassware closely for cracks, stars, and chips before use. Small flaws in the glass can lead to failure when heat is applied. Consult your instructor whenever flawed or damaged glassware is discovered, because in some cases, specialty glassware can be repaired rather than discarded. Damaged glassware should be discarded in an appropriate waste container labeled "broken glass only". Your instructor will describe the procedure to replace your glassware. Keep your workspace clear of clutter. Only glassware currently in use should be on the benchtop.

Working with Flammable Liquids or Gases

Often, many students share limited laboratory space, so be aware of other students and their experimental setup as well as their movements around the laboratory. Be sure to pay attention to the proximity of reagent bottles to open flames or other sources of ignition or heat, and ask your instructor whether your reaction setup should be relocated. Flammable solvents should be stored in a flammable cabinet if the recommended storage is at room temperature. Always minimize the quantity of solvents outside of the flammable cabinets.

Laboratory Hoods

Laboratory hoods are an **engineering control** recommended for all operations in which offensive odors or toxic or flammable vapors may be evolved. Space in the hoods may be limited, and laboratory hoods that are actively being used for experiments should not be used to store chemicals or waste. Hoods may have small glass panels that slide sideways or a larger panel that raises and lowers, or both. This part of the hood is called the sash. Once the setup portion of the experiment is completed, the sash should be lowered or closed as much as possible to maximize airflow through the working space and away from the user. The airflow in the laboratory hood can be disrupted by drafts from windows or doors and even by a change of position of the students standing at the hood. Keep the sash lowered whenever possible, and only open it the minimum amount necessary to access equipment.

Before setting up equipment in the hood, be sure that the hood is working properly. A properly operating hood requires both an adequate airflow and the absence of excessive

turbulence, so that containment of chemicals is maximized and personal exposure is eliminated. Ask your instructor if you are not sure whether the hood is operating properly. Always keep your face outside the plane of the hood sash when an experiment is in progress. When you place your equipment within the hood, it should be placed at least 15 cm (6 inches) from the front edge of the hood. Work as far back in the hood as practical, but do not block any rear exhaust openings. Whenever possible, equipment should be elevated in a laboratory hood so that the disruption of air flow is minimized.

Hoods should protect users from exposure to toxic vapors when the sash is closed, but a hood is not designed to fully protect you from physical injury if a chemical explosion occurs. When it is necessary to perform a procedure that could result in catastrophic failure (implosion or explosion), the experiment must be performed behind laboratory shields designed for that purpose. Shields should be stabilized so that they cannot be knocked over or become a projectile in case of catastrophic failure. In addition to wearing eye protection, you should use full face protection when checking the operation of a reaction that has the potential for explosion.

Distillations

A common method of separation and purification is distillation. Careful design and setup of the distillation system are required to safely accomplish effective selective evaporation and subsequent condensation of the desired product. Distillation uses differences in the volatility of a mixture's components to separate the mixture. Potential hazards include pressure buildup, heating of flammable solvents, initiation

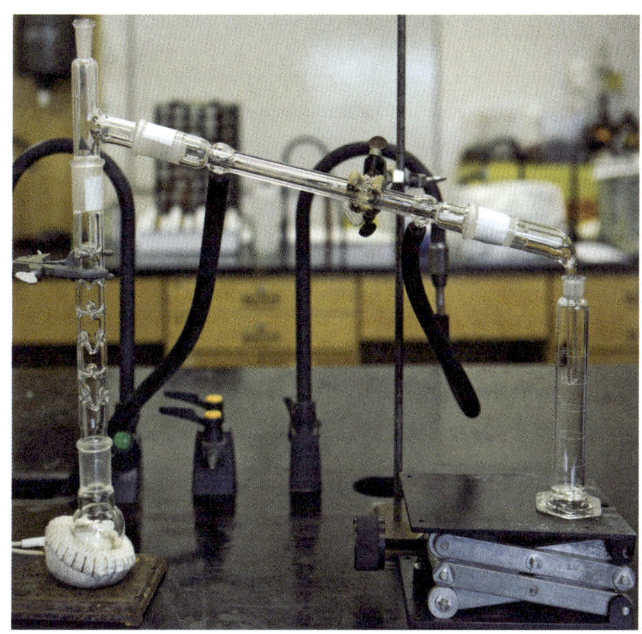

of an exothermic reaction, vapor leaks that can ignite, and continuing to heat until dryness. Carefully follow the specific directions provided by your laboratory instructor, and ask questions for clarification of the procedures if needed.

Even heating and smooth boiling are desirable to avoid bumping. Bumping refers to the superheating of a solvent, which then suddenly releases a large vapor bubble, forcing liquid up and out of the boiling flask. Your instructor may suggest that you add boiling chips to your round-bottom flask to minimize bumping. The boiling chips provide nucleation sites for the gas bubbles and prevent superheating.

Even heating with steady stirring also minimizes this concern. Your laboratory instructor may direct you to use a hot plate with a stirrer or a heating mantle. Be sure to watch the reaction temperature, and never heat above the temperature directed in the procedure.

■ IN YOUR FUTURE: Unattended Operation of Equipment

Extractions

Glass separatory funnels are commonly used to separate mixtures. Watch your instructor demonstrate the proper use, and always follow the directions provided for the separation procedure. In general, never vent the separatory funnel near a flame or other ignition source, be sure to direct the vapors away from yourself and others in the laboratory, and direct the vapors into the laboratory hood. When using a volatile solvent, first swirl the unstoppered separatory funnel to allow some solvent to vaporize and expel air. Then close the funnel, invert it with the stopper held firmly in place, and immediately open the stopcock to release air and vapor. Do this with your hand and fingers firmly holding the stopcock in place. Repeat as necessary.

Precautions for Using Electrical Equipment

Inspect the insulation on an electrical cord before use. The outlet plug should not be bent or damaged. When removing the plug from the electrical outlet, do not pull on the electrical cord. Firmly grasp the end of the plug and pull the plug directly out of the outlet. Do not remove plugs with wet hands. Damaged, tattered, or cracked cords and electrical failures or evidence of overheating should immediately be reported to your instructor. For normal power (110–115 volts alternating current) applications, only three-prong grounded, double-insulated, or isolated wiring should be used. Because of the risk of fatal shock or electrical burns, care must be taken when exposure to a live circuit could occur. Electrical currents as low as 0.1 amperes may result in fatal shock.

The most commonly used electrical equipment in your first laboratories will be hot plates and hot plates with magnetic stirring motors. When hot plates with magnetic stirrers are used, remember that the appearance of the ceramic plate does not change, whether it is cold or hot. Always carefully move the plate by grasping the bottom of the apparatus. You may burn yourself if you grasp the top ceramic plate when it is hot. Before turning the hot plate on, ensure that the cord is not in contact with the ceramic top of the hot plate. Most hot plates are not explosion-proof, and they should be used in a hood when you are heating organic solvents, so that the vapors are removed.

Note that the control for the magnetic stirrer and the temperature control often appear similar. Be sure to carefully distinguish which knob you are increasing or decreasing. Some models allow the user to continue to rotate the dial from LOW to OFF and then back to HI. Be sure to confirm that the hot plate is OFF and unplugged before you leave the laboratory.

Refrigerators

Refrigerators and freezers are used in laboratories when low-temperature storage of laboratory chemicals is required. The label on the chemical will indicate the storage temperature for the chemical. Depending on the type of laboratory, these units may be rated for storage of flammable materials. A unit rated for storage of flammable materials ("lab-safe") protects the contents within and will be clearly labeled as such on the front.

Under no circumstances should flammable substances be stored in a non-flammable-rated or household refrigerator. Chemicals stored in a refrigerator should be placed on a spill tray with sufficiently high edges to contain the entire contents if a spill occurs. NEVER, under any circumstances, store food or beverages for consumption in a refrigerator in a chemical laboratory. If an experiment requires the testing and storage of food or beverage samples, each sample should be clearly labeled "Not for human consumption".

Centrifuges

Benchtop centrifuges should be placed back from the edge of the laboratory bench, so that if excessive vibration occurs, they will not "walk" off the edge of the bench. Most modern centrifuges have suction feet that will prevent this for mild vibration. Proper loading of the centrifuge will prevent an unbalanced spin. The volume of liquid in the test tubes should be approximately the same height, and the test tubes should be placed directly across from each other or in a triangular formation (see diagram). NEVER pour liquids directly into the centrifuge tube sleeves. Modern centrifuges also have lids that prevent accidental contact with the rotor. Some even have a braking mechanism in case the lid is opened. Always close the lid before turning on the centrifuge, and keep the centrifuge closed while it is running. Allow the rotor to come to a complete stop before opening the lid. Stopping the rotor with your hands can be dangerous as well as defeating the purpose of centrifuging: to separate a precipitate from the solution. Follow the directions of your laboratory instructor, and ask for assistance if the centrifuge is not operating as expected.

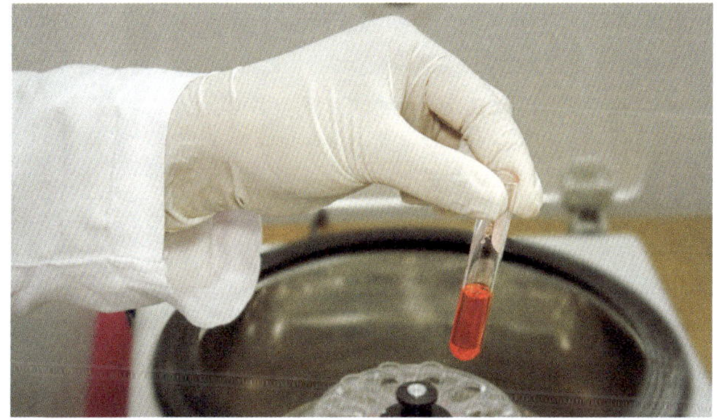

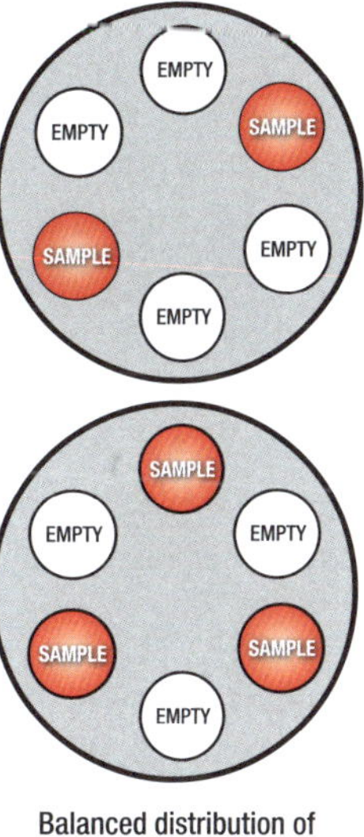

Balanced distribution of samples in centrifuge

High-Pressure Air

Many laboratories are equipped with pressurized air valves (often near the hood) for drying glassware. Only use high-pressure air from a "house system" when instructed to do so; residual oil may be present from the equipment used to generate the house air, and this can contaminate the glassware. You should never direct air under pressure toward yourself or any other person. Pressurized air can penetrate skin without any visible opening, causing the area to expand like a balloon. Severe damage to tissue can occur, which may require hospitalization.

Ultraviolet Lamps

Ultraviolet (UV) lamps are used to visualize substances that fluoresce when exposed to UV light (100–400 nm). Thin-layer chromatography (TLC) is used to separate mixtures; the spots can be observed under UV light. When using a UV lamp, care must be taken to protect your eyes. Your eyes may be accidentally exposed to UV light, so special eyewear with UV-absorbing lenses should be worn. It is preferable to operate UV lamps only in a completely closed radiation box, to minimize exposure of others around you.

General Considerations when Controlling Temperature

Many reactions require temperature control. Take note of the experimental protocol before assembling your apparatus, so that heating and/or cooling can be readily applied and/or withdrawn to manage temperature. Carefully follow the directions of your instructor whenever an exothermic reaction is expected. Often, chemical reactions must be initiated by heating. In an exothermic reaction, temperature increases. The rate at which the heat is produced increases exponentially, whereas the rate at which the heat is removed is linear. Highly exothermic reactions can very quickly become dangerously violent, in what is called a thermal runaway reaction, unless adequate cooling can be readily applied.

Some exothermic reactions proceed with an initial slow phase followed by a rapid acceleration. The initial slow phase is known as the induction period. In such reactions, if too much reagent was initially added, the reaction can become too vigorous for effective condensation of vapors once the induction period is completed. Once the rapid phase of a runaway reaction ensues, external cooling must be used to control temperature. A cooling bath must be prepared in advance and be at the ready so that it can be promptly applied under the reaction vessel.

Do not heat closed systems unless specifically instructed to do so by the procedure or by your laboratory instructor. Consider the need for the buildup of excess pressure to be released when an apparatus is heated or an exothermic process is anticipated, and have a mechanism in place to vent excess pressure.

When temperatures lower than those that can be attained with an ice bath are required, dry ice or dry ice combined with an organic liquid may be used. Dry ice will sublime at −78 °C; a mixture of dry ice with acetone maintains this temperature, and the acetone will not freeze. Liquid nitrogen (LN$_2$, −196 °C) must be used with extreme care and only when you are instructed to do so by your instructor or research advisor. In addition to proper laboratory attire, chemical splash goggles and insulated cryogenic gloves are required when working with these very cold baths. Dry ice and LN$_2$ are also both simple asphyxiants and must be handled in a well-ventilated room or in a laboratory hood.

Cooling Baths

When ice water is not sufficiently cold for use as a bath or trap, salt and ice may be used. Depending on the number of moles of ions (recall the concept of freezing-point depression), the ice baths can be prepared in the range of −10 °C to −40 °C.

Determination of Melting Point

The accurate determination of the melting point of a compound is a common method of identification. In the past, a melting-point apparatus may have used a mercury thermometer, because they are more accurate than alcohol thermometers. Because of the hazards associated with breakage of a mercury thermometer, most laboratories now use an electrothermal melting-point apparatus. The main hazard here is the very hot metal chamber that holds the sample.

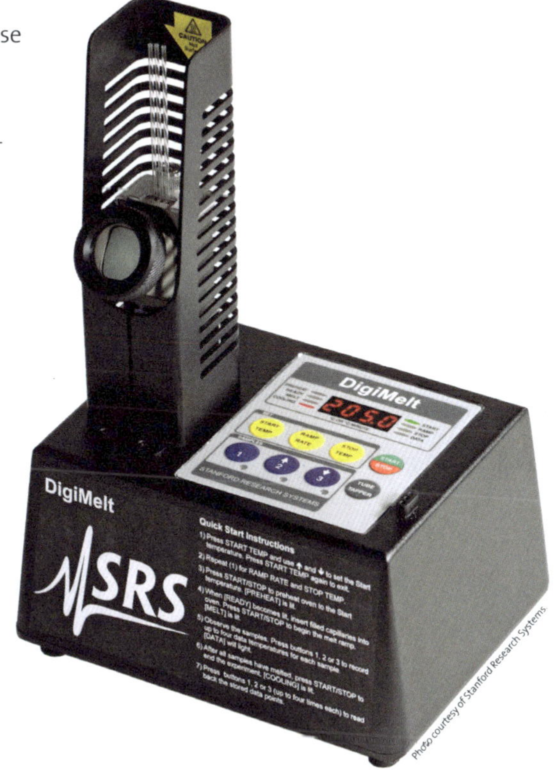

Use of Bunsen Burners

Bunsen burners may be used in laboratories to heat samples strongly or to boil water rapidly. A well-adjusted flame will have two distinct cones: an outer aqua cone and an inner blue cone. The temperature of the hottest part of the flame (the tip of the inner blue cone) is about 1200 °C. Take a moment to look at the burner. You will see a barrel with a removable tip at one end and a band of holes at the other end. These holes adjust the amount of air entering the barrel (a mixing chamber for fuel and air). On the bottom is a needle valve, which, when turned tight into the barrel, blocks all gas from entering the barrel. Before using a Bunsen burner, check the hose for cracks, and ask your instructor to replace a faulty hose. Your instructor will show you how to adjust the air holes, needle valve, and gas jet to light the burner with a laboratory striker. Before using a flame, tie back loose hair and remove or confine scarves.

Heating a Test Tube

When the contents of a test tube are heated in an open flame, the contents can easily be overheated, causing the contents to boil up and forcefully eject out of the test tube. Always hold the open test tube at an angle facing away from yourself and others. While holding the test tube with a test tube holder at an angle, heat it gently along the side, not at the bottom of the tube. An alternative is to heat the contents of a test tube by placing it in a hot water bath.

Reduced Pressure

Water aspirators are sometimes used to create reduced pressure for filtration. Use only heavy-walled filter flasks that are designed for this purpose; do not apply reduced pressure to other flat-bottom flasks, which may not withstand reduced pressure and could crack during use. The flask should be clamped to a ring stand for stability. A trap is placed between the aspirator and the filtration flask, so that water cannot be sucked back into

your filtrate if the water pressure decreases unexpectedly and so that the filtrate cannot enter the sink. Whether or not you are using a trap, always disconnect the vacuum hose line on the side arm before shutting off the water source. More and more, the use of water aspirators is generally discouraged, because of the significant quantities of wastewater produced.

SUMMARY

In this chapter, the safe use of common laboratory equipment was reviewed. Often, laboratory processes require pressures and/or temperatures other than ambient, and therefore can introduce physical hazards into your experiments. Always review the laboratory directions before beginning any procedure. Continually assess your situation in the laboratory and be aware of your surroundings, and you will learn to proactively recognize potential hazards.

Ask your laboratory instructor if you have any concerns about the safe use of equipment and the completion of any procedures. Chapter 5 presents protocols that will help you prevent and prepare to respond to laboratory incidents.

REFERENCES

[1] ACS Green Chemistry Institute. www.acs.org/content/acs/en/greenchemistry.html (accessed March 6, 2017).

Safety Equipment and Emergency Response

Introduction

Although the laboratory is designed to be a place for learning and skill development, it is important to prepare to respond appropriately to unexpected events. Even when recognized hazards are appropriately managed, incidents may occur. In an academic setting, the instructor's responsibility to respond to an incident is certainly greater than the student's, but the intent of this chapter is to help you prepare for the unexpected as a student and later as a scientist. With this in mind, always follow the directions of your instructor and the policies of your institution.

In previous chapters, you have learned about the first three principles of RAMP: Recognize the hazards, Assess the risks of hazards, and Minimize the risks of hazards. The first three principles of RAMP will reduce incidents and help to avoid emergencies but may not prevent them entirely. The fourth part of RAMP is Prepare for emergencies.

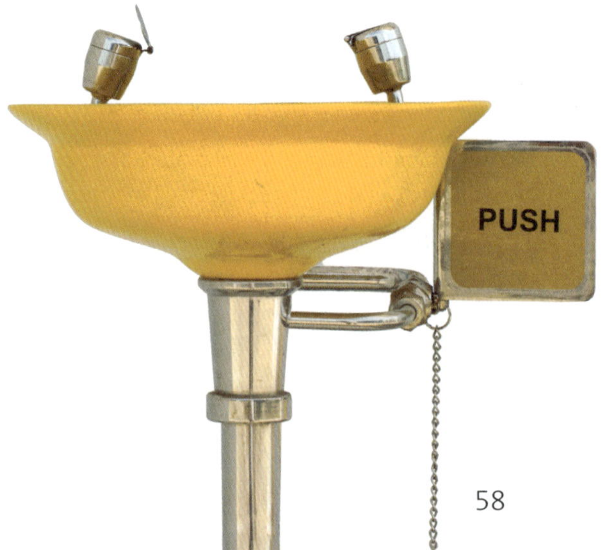

This chapter covers how to ***prevent incidents from occurring and prepare to respond*** to unexpected emergencies. You are more likely to remain calm and respond appropriately to an emergency if you have put some effort into thinking about potential incidents and your response. In introductory laboratories, the most common emergencies that you may need to respond to involve spills, fires, cuts, and/or burns.

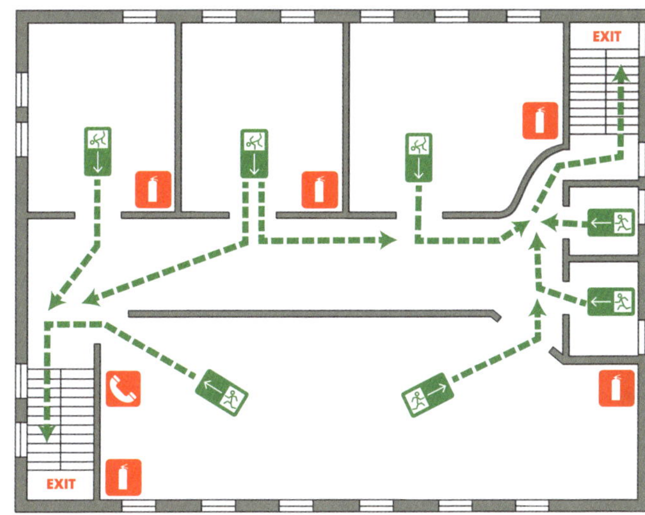

As you begin to think about how you might respond to an emergency, take a moment to look around the laboratory when you first enter to identify safety equipment, signs, fire alarms, and exits. Some of the items you can locate include a fire extinguisher, safety shower, eyewash station, evacuation signage, emergency gas shutoff (if accessible), first aid kit, and fire blanket. If your laboratory is equipped with more than one of each of these items, note the location of those closest to you. Some of this safety equipment, such as fire extinguishers and eyewash stations, is tested periodically and will have a tag including the date that testing is due. Although it is not your responsibility to inspect this safety equipment, notify your instructor if you notice that any of the equipment is overdue for inspection or if access is blocked.

It is likely that your instructor will be the first to respond to a laboratory incident. However, your instructor may be momentarily away, working with another student, or farther away from the incident, and you will need to respond quickly. While preparing to respond to an emergency, remember that the primary goal is to protect human life and minimize injury. Never put yourself in danger. The second goal is to minimize damage to structures or equipment. You should follow the emergency procedures established by your institution under the direction of your instructor.

In certain situations, you may be the only person who can respond. Before you attempt to help another person, assess the current situation and evaluate the potential risks to yourself. You cannot help another person if you are injured in the process. In general, the emergency

These terms are often used inter-changeably in everyday conversations. Many dictionaries use these terms to define each other. The following is typical (and not very useful): "Combustible: flammable; capable of combustion."

When these terms are used to describe the fire risk associated with organic solvents, the National Fire Protection Association (NFPA) and the Globally Harmonized System of Classification and Labelling of Chemicals (GHS) use two criteria that affect the fire risk present when using a solvent: the flash point and the boiling point. The flash point of a volatile solvent is the minimum temperature at which sufficient vapor is given off from the liquid surface to form an ignitable mixture with air.

Unfortunately, the NFPA and the GHS define slightly different criteria for "flammable" and "combustible", but the general difference is that flammable liquids ignite and burn at normal temperatures and combustible liquids burn at higher temperatures. Flammable liquids generally present a greater hazard in the laboratory. Combustible liquids may or may not continue to burn when the ignition source is removed.

In both the NFPA system and the GHS, there are additional criteria defining these terms for gases and solids. In all cases, combustible and flammable materials will add fuel to a fire, and appropriate precautions must be taken in the laboratory.

procedure that you may need to follow will depend on the type of immediate danger and injuries sustained, but may include elements of the following steps.

- Determine whether you and others need to leave the area immediately or whether there is something you can safely do to minimize injury and damage (see later sections).

- Report the nature and location of the emergency to your instructor or, if necessary, call 911 (or your institution's emergency number) to contact the appropriate fire or medical assistance. Be prepared to answer questions from the dispatcher, such as your name, the location, and the nature of the incident. It may be necessary to send someone to meet the ambulance or fire crew at the building entrance, because these crews may not be familiar with your building.

- Notify others in the area about the nature of the emergency, and call the security office on your campus.

- Do not move injured individuals unless they are in immediate danger from chemical exposure or fire. Keep them warm. Unnecessary movement can severely complicate neck injuries and fractures.

Fires

Fire Prevention

The best way to fight a fire is to prevent it. Think about fire prevention using the principles of RAMP.

- Are you working with any source of heat, flame, or spark?
- Are you working with flammable liquids or vapors?
- Are there any damaged wires on the electrical equipment?
- Are bottles or glassware (containing flammable solvents) too close to the edge of the laboratory bench?
- Is the workspace cluttered?

When you are using a flammable liquid, minimize the quantity in your workspace by dispensing only the quantity required by the experimental procedure and returning the reagent bottle to its proper storage location immediately after dispensing the material. Excess flammable material in the immediate work area will act as an additional fuel source in a fire. You should also keep combustible materials, such as paper, away from areas where experiments are being performed with flammable chemicals.

In the event of a fire, combustible materials will also act as fuel to keep the fire burning. NEVER store combustible materials on top of a flammable cabinet.

When heating flammable solvents, it is important to avoid the presence of ignition sources. An open flame should never be used for heating a flammable solvent. Electrical hot plates are safer, but electrical equipment can often generate a small spark when it cycles on and off. This spark can start a fire if sufficient vapor from a solvent is present. Some electrical equipment is

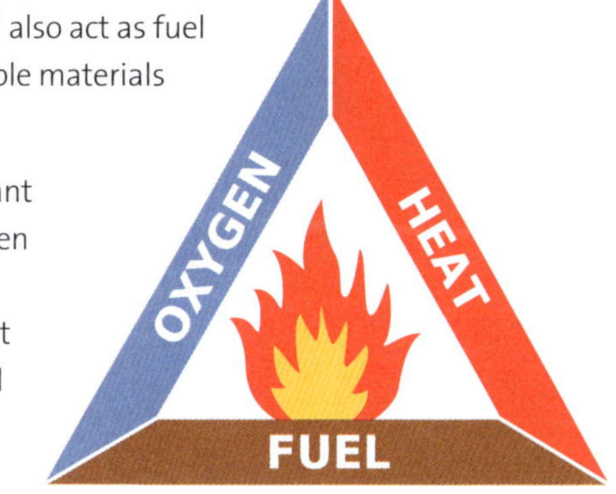

designed to be "intrinsically safe", which means that it is designed to prevent such sparks, but this is not the common situation in most laboratories. Thus, it is important to avoid the possibility of flammable vapors when heating solvents, which generally requires that this process be done in a hood. If any electrical equipment is sparking or has frayed wires, turn it off immediately and inform your instructor.

Good housekeeping habits will go a long way to preventing incidents and minimizing the effects if an incident occurs. Although it is primarily the instructor's role to regularly inspect electrical cords on equipment and to make sure that aisles and exits are clear, learning to survey the "scene" or area is a good habit for you to develop. Even when risk is managed well, an emergency situation may still arise in an academic laboratory with many students present. It may be that you find yourself responding to a situation that you did not cause.

Prepare to Respond to a Fire

Consider the three components of the fire triangle: heat, fuel (flammable or combustible material), and oxygen (or other oxidizer). All three components must be present for a fire to begin and continue. Removing one of these components will prevent the fire or extinguish one that has started.

If a fire unexpectedly occurs, it could be that a quick response on your part will prevent a small incident from becoming a large incident. You will need to decide whether you should fight the fire or flee to safety, but always follow the directions of your instructor. If you decide to try to put out the fire, you will want to be confident that your actions will not spread the fire or put yourself or others in more danger. It is best practice for your institution to conduct its own fire training as part of the safety training associated with the laboratory. If so, you and your fellow students should know what is expected in a fire emergency and follow these guidelines.

You should never pick up a vessel or piece of equipment that is on fire. The physical act of throwing water on some types of fires may spread the fire. The best method of extinguishing a small laboratory fire is to smother it. Consider taking the following steps only if you are confident that you can do so safely as you prepare to respond to a fire emergency.

■ Classes of Fires

The National Fire Protection Association (NFPA) identifies four main types of fires:

Class A: *fires involving ordinary combustibles, such as paper, wood, and furniture;*

Class B: *fires involving flammable organic liquids and gases;*

Class C: *fires involving live electricity; and*

Class D: *fires involving active metals (which usually react with water).*

Most fires are Class A fires. In some chemistry laboratories where organic solvents are used, Class B fires are possible, and common. Some Class A fires are also Class C fires, such as a burning computer. The class of a fire helps to determine what kind of extinguisher to use; in fact, fire extinguishers are commonly labeled as "A", "BC", or "ABC". The fire code requires that portable extinguishers that are made available in hallways (or in laboratories) are the proper type for the most likely kind of fire. Most laboratories have "ABC" extinguishers.

Class D fires are uncommon in most offices and buildings, but they are a possibility in laboratories where active metals are used. Therefore, a Class D extinguisher should be available in laboratories where Class D fires are possible.

- Remove the heat source by turning off the gas supply to a Bunsen burner or unplugging an electric heater.

- A fire contained in a small vessel often can be suffocated by limiting the air/oxygen. For example, a fire in a small beaker or Erlenmeyer flask can be smothered using a noncombustible item, such as a watch glass held in tongs or the flat bottom of a larger beaker.

- NEVER use dry towels or cloths to cover the fire, because these will burn to add fuel. You can use a wetted material if available. To avoid spreading the fire, remove nearby combustible or flammable materials, but only if it is safe to do so.

- If the steps listed above are not practical or do not extinguish the fire, a fire extinguisher may possibly be used to extinguish the fire. If you have been trained in the use of a fire extinguisher or otherwise feel confident using one, position yourself between the fire and an escape route (e.g., a door), and fight the fire from this position to be sure that you can escape. Small fires often can be extinguished, but not always. It is easy to underestimate a fire. If not extinguished, a fire can quickly threaten your life.

- If the fire is burning over an area too large for the fire to be extinguished quickly and simply, everyone should evacuate the area. Pull the fire alarm, and follow the evacuation procedures established by your institution. Once you are safe, make sure that you or someone else calls 911 (or your institution's emergency number) to notify the proper emergency assistance. Although it is difficult to know who is inside a university building when a fire breaks out, it is important to locate other students and faculty outside of the building to let them know that you are safe. Offer assistance to others who may be mobility-impaired, if you are able.

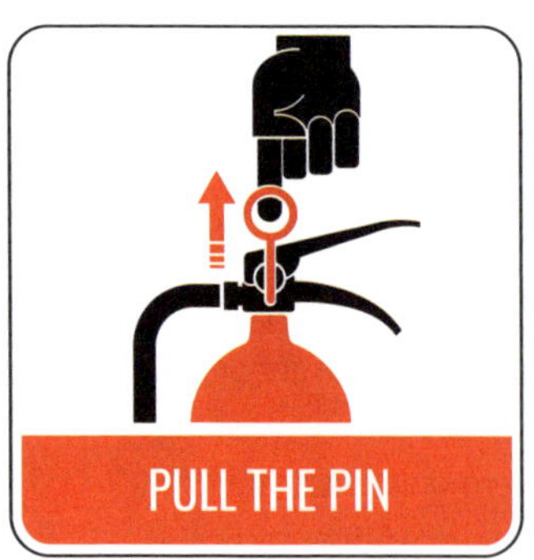

When You Are On Fire

1 STOP 2 DROP 3 ROLL

Prepare to Respond to Personal Injuries Involving Fires

If someone's clothing catches on fire, this will be a frightening situation, and the person involved may instinctively run. Running will make the situation worse by fanning the flames, and the heat and gaseous product of combustion are likely to be inhaled for someone who is standing, walking, or running. Firmly tell the person to **stop, drop, and roll**. This may smother the flames. You can try to extinguish any small, still-burning flames by patting them out with a towel or jacket. After getting the person to lie down, start at the head and the shoulders, and then work downward toward the feet. These recommendations follow the procedures outlined by the National Fire Protection Association (NFPA), which assume that someone is in an ordinary, non-laboratory environment. Another option in the laboratory is to douse the person under a safety shower or with water from a drench hose at a laboratory sink, if these are convenient options. You will have to make a judgment call about the best way that you can help.

Never wrap a standing person in a fire blanket, because this may direct flames and toxic gases to the face area in what is known as the chimney effect. Fire blankets can be used as a modesty curtain while an individual is in the shower, to keep an injured person warm, as a pad to keep the person off the floor, as a wrap if no emergency clothing is available, and as a stretcher. If your institution has fire blankets, you should follow local instructions about how to use them. Get medical attention promptly by calling 911 (or your institution's emergency number). Covering the victim with a blanket or jacket may help to avoid shock and exposure.

Chemical Contamination on Skin, Clothing, and Eyes

Preventing Chemical Contact

Exposure to chemicals can arise from incidental contact with the skin, splashes, and contact with spills. Spills, and how to prevent and respond to them, are discussed below. Chapter 2 covers practical rules for the laboratory to help you to avoid chemical contact, including various forms of personal protective equipment (PPE). You may want to review those practical rules, which can prevent incidents.

Prepare to Respond to Chemical Contact

Once again, your instructor will likely be present to assist you or another student in the event of an incident. The advice given here is general and will help you to know how to respond to an incident.

For a small liquid spill or splash that affects only a small area of skin, you should immediately flush the skin with flowing water for at least 15 minutes (30 minutes for bases). You should remove any jewelry, to facilitate removal of possible residual liquid between your skin and the jewelry. This is one reason why you are advised not to wear jewelry in the laboratory at all. After the initial rinse with water, wash the entire area of the affected skin with warm water and soap. You or a fellow student should notify your instructor as soon as possible. After rinsing your skin, check the Safety Data Sheet (SDS) to see whether any delayed effects should be expected. It is advisable to seek medical attention for even minor chemical burns. Many institutions will require you to fill out a short initial notification of injury form. This is for your protection should complications develop from an exposure in the laboratory.

If you spill solid chemicals on your skin, it is advisable to brush the solid off before applying water. (A credit card is useful for this procedure.) Brushing off a corrosive solid can usually be done with no adverse consequences, and may actually lessen the exposure by avoiding increased temperature due to an exothermic heat of solution, which is common when water is added. After the material is brushed away from your skin, wash your skin with soap and water and notify your instructor of the incident. The brushed-off solid should, of course, be put into the appropriate waste container, rather than down the sink or into the regular trash. In the event of contact with an acid, do not put a neutralizing chemical (such as sodium bicarbonate) on your skin. The heat of neutralization can increase an injury.

If your skin or clothing is contaminated with larger spills of a liquid, you may have more serious consequences. You should go to the nearest safety shower immediately, turn on the water, and quickly step under the water spray. Some showers will remain on until intentionally turned off, but others may require that you hold the handle in the "on" position to keep water flowing. While you are under the flowing water, remove all of your contaminated clothing, shoes, and jewelry. Seconds count, so don't waste time with modesty. Delays due to modesty have resulted in exposures turning into serious injury. Avoid contaminating your eyes and face by allowing someone else to cut off a pullover shirt or sweater with scissors. You should flood the affected body area with water for at least 15 minutes, but resume flushing the area with water if pain returns. It is unlikely that your instructor would not be aware of a situation requiring the use of a safety shower, but notify your instructor as soon as possible. After thoroughly flushing the area with water, seek medical attention. If possible, your instructor will take the SDS to the medical facility, but you may be helpful in identifying the name of the chemical. Many institutions will have extra clothing (such as sweatshirts and pants, or hospital scrubs) to immediately replace your contaminated clothing. As indicated in Chapter 2, wash contaminated clothing separately from other clothing or discard it, as recommended in the SDS.

If you are splashed in the eye with chemicals, seconds count. Quickly go to the nearest eyewash station, and flush the eye for at least 15 minutes (30 minutes for bases). You should use your thumb and forefinger to hold your eyelids away from the eyeball, and move your eye continuously — up and down and sideways — to flush the area behind the eyelid. If you are assisting someone else who has had chemicals splashed into their eyes, you may need to lead them to the eyewash station. It is worth repeating here that wearing chemical splash goggles will essentially eliminate any chance of this kind of incident occurring.

If an eyewash fountain is not available, you should use the nearest source of running water and make the water as close to room temperature as you can. Alternatively, you could help an injured person onto their back and pour water gently into the corners of the affected eye for at least 15 minutes. If possible, remove contact lenses. Your instructor should be notified as soon as possible. After flushing the affected eye, seek medical attention. If possible, identify the chemical contaminant, so that your instructor can take a copy of the SDS to the medical facility.

NOTE: Although current best practices call for tepid water to be dispensed from eyewash fountains and safety showers, this is likely not the case everywhere. Should the exposed person start shivering or complain about being too cold, use your best judgment about the rinsing time. You do not want to induce hypothermia in a person with a chemical exposure.

As mentioned in Chapter 3, the hazards of hydrofluoric acid (HF) make its use very high-risk. HF exposures are extremely hazardous and require special treatment; a calcium gluconate gel or cream should always be available when using HF. If skin is contaminated with HF, you should apply this gel and immediately seek emergency medical treatment. You should not work with HF without special training and supervision.

Other Personal Injury

Preventing Other Personal Injuries

Other types of personal injuries include slips, trips, falls, electrical shock, chemical ingestion, inhalation, and cuts. You should review the general rules in Chapter 2 and the use of chemical hoods and electrical equipment in Chapter 4 to avoid these incidents.

Prepare to Respond to Other Personal Injury Incidents

Although fires and spills are the most common types of laboratory incidents, you should think about how you would respond to some other possibilities, including chemical inhalation, chemical ingestion, electrical shock, and cuts.

If you or someone else in the laboratory is overcome with smoke, vapors, or fumes, move yourself and others away from the area to fresh air. You should warn other people in the area of the potential for harm, and seek medical assistance immediately. If you suspect that someone has ingested hazardous chemicals, call 911 (or your institution's emergency number) and follow the first aid treatment shown on the label or in the SDS. You may be asked to assist your instructor with first aid treatment. You should never give anything by mouth to an unconscious person. Your instructor will ensure that the medical facility is provided with a copy of the SDS, but you may be needed to identify which chemical is involved.

You should not touch someone who is in contact with a live electrical circuit. The circuit must be disconnected by unplugging the device or turning off the circuit breaker, or you will be shocked too.

If the injured person is not breathing and has no pulse, you should provide cardiopulmonary resuscitation (CPR) if you are trained to do so, or use an automated external defibrillator (AED) if one is available. You should call 911 (or your institution's emergency number) immediately, or tell someone else to do so while you are tending to the victim. Because it is likely in academic laboratories that several people will be present, it is best to have one person call 911 and another person locate a laboratory instructor or other person in authority, call campus security, and assist with CPR.

If you or someone else is bleeding severely, try to control the bleeding by placing a cloth on the wound and applying firm pressure. If possible, elevate the injury above the level of the heart. **You should take reasonable precautions to avoid contact with someone else's blood.** There are gloves available in most chemistry laboratories. Except in the case of a trivial wound, wrap the injured person in a blanket or coat to avoid shock and get them immediate medical attention. **Do not clean up a biological spill that may contain blood-borne pathogens unless you are trained to do so.**

Chemical Spills

Preventing Chemical Spills

Chemical spills are probably the most common laboratory incidents. Good housekeeping habits will help you to avoid spilling chemicals. Keep laboratory items well away from the edge of your laboratory bench or other workspace. You should measure quantities of chemicals according to the laboratory procedure and avoid taking excess chemicals. Return reagent bottles to their proper location once you have procured the minimum amount needed for your experiment. Notify your instructor if you observe obstructions in the aisles or walkways of the laboratory space. Walk slowly and carefully in the laboratory. Rushing may cause you to bump into other students or into cabinets and laboratory furnishings.

If you must transport samples or solutions from one part of the laboratory to another, support the beaker or flask with one hand under the container. Alert others to your presence if needed. The transport of chemicals out of the laboratory to an instrument or storage area requires secondary containment with a rubber carry bucket or plastic tote.

Prepare to Respond to a Chemical Spill

If you or someone near your workspace has a chemical spill, you and other students in the area should move away from the spill. If a flammable liquid is spilled, warn other students in the area to extinguish all flames and turn off electrical equipment, if you can do so without putting yourself in harm's way. If the spill occurs in a chemical hood, close the hood sash to allow the vapors to be removed more effectively. You should report the spill to your instructor immediately. Your instructor should know how to handle the spill. If a significant amount of flammable, toxic, or volatile material is spilled, it may be necessary to vacate the entire laboratory or building.

If a small spill of a solid material occurs, your instructor may direct you or others to use a dustpan and brush, typically stored in the laboratory, to clean up the material. If a small spill of a liquid material occurs, your instructor may direct you or others to use paper towels or another absorbent to soak up the liquid. To pick up broken glass, use tongs or wear leather or cut-resistant gloves. Broken glass can also be swept up using a small brush and dustpan. Most likely, an instructor will clean up broken glass, or you may be asked to do this under close supervision. Dispose of the material, including paper towels, as directed by the instructor.

If a large spill of a material occurs, or if the material is toxic or flammable, your institution will have a formal procedure for handling the spill. Follow the directions given by your instructor. These spills are not handled by students in introductory laboratories.

Your instructor or other trained personnel may contain larger liquid spills on the floor by surrounding the involved area with an absorbent retaining material. The absorbent material used may be one that neutralizes the spilled material (limestone or sodium carbonate for acids, sodium thiosulfate solution for bromine, etc.). There are commercial absorbent kits (e.g., Oil-Dri and Zorb-All), but other readily available absorbents, such as vermiculite or small particles (about 30 mesh) of *clay-based* kitty litter, can also be used effectively. When you are directed to return to the area, make sure that the floor does not feel slippery to you. Let the instructor know if the area seems unsafe for any reason.

SUMMARY

The most likely emergency situations that might arise in an introductory or organic chemistry laboratory are ignition of a flammable solvent, chemical exposure on skin or in eyes, cuts from broken glassware, and chemical spills. You are more likely to respond appropriately, thus minimizing injury and damage, if you have considered your response in advance.

Learning to work safely in a laboratory is an important component of undergraduate chemical education. Recognizing hazards and assessing and minimizing risks are habits that should continue to be developed throughout your education and your career. This chapter has presented information on how to **prevent an incident or spill**. Prevention education is important, but **preparing to respond** in an emergency is just as important and should also be part of your education.

APPENDIX
The Web as a Source of Safety Information

The Internet offers a multitude of resources on chemical safety. Many of these websites contain a mixture of accurate and inaccurate information. A few are outright unreliable, being little more than expressions of ill-founded opinions concerning chemical safety matters and the environment. Below are some recommended websites where accurate information can be found.

Recommended Websites

The American Chemical Society offers authoritative information about chemical safety on various websites.

- The ACS Committee on Chemical Safety (CCS) facilitates the promotion of chemical safety in various ways, such as the publication of this booklet. The CCS website offers numerous resources to assist instructors and students in teaching and practicing chemistry safely. **www.acs.org/content/acs/en/about/governance/committees/chemicalsafety.html**

- The ACS Division of Chemical Health and Safety (CHAS) website offers emerging information on all aspects of chemical and laboratory safety. **www.dchas.org**

Other Useful Sites

- The U.S. Occupational Safety and Health Administration (OSHA) has information relevant to all aspects of worker safety, including statistics, a description of the agency, its newsroom (speeches, news releases, testimony, and publications), and OSHA regulations. **www.osha.gov**
  Of special interest are Standards in 29 CFR (indicating that they were published in Volume 29 of the Code of Federal Regulations). Section 1910.1450, titled "Occupational exposure to hazardous chemicals in laboratories", is the "Laboratory Standard", and it is of particular relevance to those who work in academic laboratories. **www.osha.gov/pls/oshaweb/owadisp.show_document?p_table=STANDARDS&p_id=10106**
  Also, at OSHA's "Safety and Health Topics" webpages, you can find subsequent links for more specific information about chemical safety, including a link to an alphabetical list of a wide variety of topics at OSHA **www.osha.gov/SLTC/index.html** **www.osha.gov/SLTC/text_index.html**
  Most of these items are unrelated to chemical safety, but the list can allow a more targeted search of a specific chemical or chemistry-related topic.

- A Guide to the Globally Harmonized System of Classification and Labelling of Chemicals (GHS) explains the GHS in a condensed version. **www.osha.gov/dsg/hazcom/ghsguideoct05.pdf**

- The U.S. Environmental Protection Agency has information about hazardous waste and environmental issues. At the "Science and Technology" page, you can see a list of subtopics of possible interest, including the website about hazardous wastes and waste management.
  www.epa.gov
  www.epa.gov/science-and-technology
  At "Laws and Regulations", you can find more links to subtopics of interest to chemists. Of particular interest to chemists interested in green chemistry and sustainability chemistry are the links at the "Sustainable Practices Science" webpage.
  www.epa.gov/laws-regulations
  www.epa.gov/science-and-technology/sustainable-practices-science

- The National Institute for Occupational Safety and Health (NIOSH) provides much information, including the NIOSH Pocket Guide to Chemical Hazards, which gives incompatibility information on about 700 chemicals. This publication can be read online or ordered for free from NIOSH.
  www.cdc.gov/niosh/homepage.html
  www.cdc.gov/niosh/npg

- The National Library of Medicine is part of TOXNET and is home to many of the most authoritative databases of specific information about chemicals, such as ChemID*plus* Advanced, PubMed, and the Hazardous Substances Data Bank.
  www.nlm.nih.gov

- Safety Data Sheets (SDSs) can give you important information about the hazards of a chemical. The SDS from the manufacturer should be used for a purchased chemical. For general information, MilliporeSigma (formally Sigma-Aldrich, has well-developed SDSs that anyone can browse online. Simply type a chemical name or a Chemical Abstracts Service (CAS) Registry Number into the main search area at the top of the homepage and select the SDS link for a product.
  www.sigmaaldrich.com/united-states.html
  If you need information on how to interpret an SDS, try the MS-Demystifier.
  www.ilpi.com/msds/ref/demystify.html

- *Prudent Practices in the Laboratory: Handling and Management of Chemical Hazards*, Updated Version (2011), published by the National Academies Press, is free to download or read online.
  www.nap.edu/read/12654/chapter/1

- The U.S. Coast Guard compatibility chart and storage codes found in Chapter 7 of *The Chemical Hazards Response Information System (CHRIS) Manual* give incompatibilities based on broad chemical classes.
  www.uscg.mil/hq/nsfweb/foscr/ASTFOSCRSeminar/References/CHRISManualIntro.pdf

- The National Oceanic and Atmospheric Administration (NOAA) database known as CAMEO (Computer-Aided Management of Emergency Operations) Chemicals offers the ability to "mix" chemicals "in vitro" to predict reactivity. The software can be downloaded or used online.
  www.cameochemicals.noaa.gov

Many other reputable websites containing safety information exist. For example, most universities and colleges as well as various other public-spirited associations and organizations maintain websites. Until you have developed a strong knowledge base, you may wish to rely on personal recommendations from your safety professionals and instructors for help in evaluating the reliability of Internet resources.

NOTES